Studies in Autonomic, Data-driven and Industrial Computing

The book series Studies in Autonomic, Data-driven and Industrial Computing (SADIC) aims at bringing together valuable and novel scientific contributions that address new theories and their real world applications related to autonomic, data-driven, and industrial computing. The area of research covered in the series includes theory and applications of parallel computing, cyber trust and security, grid computing, optical computing, distributed sensor networks, bioinformatics, fuzzy computing and uncertainty quantification, neurocomputing and deep learning, smart grids, data-driven power engineering, smart home informatics, machine learning, mobile computing, internet of things, privacy preserving computation, big data analytics, cloud computing, blockchain and edge computing, data-driven green computing, symbolic computing, swarm intelligence and evolutionary computing, intelligent systems for industry 4.0, as well as other pertinent methods for autonomic, data-driven, and industrial computing.

The series will publish monographs, edited volumes, textbooks and proceedings of important conferences, symposia and meetings in the field of autonomic, data-driven and industrial computing.

More information about this series at http://www.springer.com/series/16624

Jyotsna Kumar Mandal · Somnath Mukhopadhyay ·
Aynur Unal · Santanu Kumar Sen
Editors

Proceedings of International Conference on Innovations in Software Architecture and Computational Systems

ISACS 2021

Springer

Editors
Jyotsna Kumar Mandal
University of Kalyani
Kalyani, West Bengal, India

Somnath Mukhopadhyay
Assam University
Silchar, Assam, India

Aynur Unal
Stanford University
Palo Alto, CA, USA

Santanu Kumar Sen
Guru Nanak Institute of Technology
Kolkata, West Bengal, India

ISSN 2730-6437 ISSN 2730-6445 (electronic)
Studies in Autonomic, Data-driven and Industrial Computing
ISBN 978-981-16-4300-2 ISBN 978-981-16-4301-9 (eBook)
https://doi.org/10.1007/978-981-16-4301-9

This Springer imprint is published by the registered company Springer Nature Singapore Pte Ltd.
The registered company address is: 152 Beach Road, #21-01/04 Gateway East, Singapore 189721, Singapore

Preface

It is a matter of great happiness for us as the editors to ventilate some of our experience in the form of an editorial preface in relation to the First International Conference Proceedings on "Innovations in Software Architecture and Computational Systems (ISACS 2021)." What indeed is an instance of pride is the articles which are eventually appearing in the scope of this proceedings qualifying all the stringent technical filter mechanisms befitting the conference proceedings published in the Studies in Autonomic, Data-driven and Industrial Computing series published by Springer Nature. It is needless to mention that compliance with all the requirements as laid down in the publication policy is a challenging task which the authors of these papers did fulfill. No doubt it is an admirable achievement on their part. As the editors of the proceedings, we desire to avail of this opportunity to express our heartfelt congratulations to them all.

We sincerely appreciate that our call for papers was able to attract forty-four submissions out of which eighteen could stand the rigorous test for being qualified for inclusion in the proceedings despite all these. We feel particularly moved because of the submissions from the prospective authors, the associated uncertainties notwithstanding.

Coming to the composition of the articles appearing in the proceedings, we categorized them broadly into four domains, viz. (i) intelligent and hybrid system, (ii) intelligent software architecture, (iii) machine learning and analysis, and (iv) sensors and smart applications. It is needless to mention that all these articles are highly commendable in a technical sense.

We were particularly impressed the way Springer Nature, as our publishing partner, did consider a stand to ensure the quality of the proceedings. Last but not least, the success of a special issue like this would not have been possible without the active participation and support extended by the authors and the learned reviewers in their respective roles.

Kalyani, India — Jyotsna Kumar Mandal
Silchar, India — Somnath Mukhopadhyay
Sodepur, India — Santanu Kumar Sen

Contents

About the Editors

Jyotsna Kumar Mandal, M.Tech. in Computer Science from the University of Calcutta in 1987, awarded Ph.D. (Engineering) in Computer Science and Engineering by Jadavpur University in 2000. He is working as a professor of Computer Science and Engineering and the former dean, Faculty of Engineering, Technology and Management, KU, for two consecutive terms during 2008–2012. He is the director of IQAC, Kalyani University, and the chairman, CIRM, and placement cell. He served as a professor of Computer Applications, KGEC, as an associate professor of Computer Science, an assistant professor of Computer Science, North Bengal University for fifteen years, and as a lecturer at NERIST, Itanagar, for one year and has 34 years of teaching and research experience in coding theory, data and network security and authentication, remote sensing and GIS-based applications, data compression, error correction, visual cryptography and steganography. He was awarded 23 Ph.D. Degrees, one submitted and 8 are pursuing; supervised 03 M.Phil., more than 70 M.Tech. and more than 125 MCA dissertations; a guest editor of MST Journal (SCI indexed) of Springer; published more than 400 research articles out of which 180 articles in International Journals; published seven books from LAP Germany, IGI Global, Springer, etc.; organized more than 40 International Conferences and Corresponding Editors of edited volumes and conference publications of Springer, IEEE, Elsevier etc.; and edited more than 40 volumes as volume editor. He received "SikshaRatna" award from Higher Education, Government of West Bengal, India, in the year 2018 for outstanding teaching activities; Vidyasagar award from International Society for Science Technology and management in the fifth International Conference on Computing, Communication and Sensor Network; Chapter Patron Award, CSI Kolkata Chapter on 2014, "Bharat Jyoti Award" for meritorious services, outstanding performances and remarkable role in the field of Computer Science and Engineering on 29 August 2012 from International Friendship Society (IIFS), New Delhi; and A. M. Bose Memorial Silver medal and Kali Prasanna Dasgupta Memorial Silver medal from Jadavpur University.

Dr. Somnath Mukhopadhyay is currently an assistant professor at the Department of Computer Science and Engineering, Assam University, Silchar, India. He completed his M.Tech. and Ph.D. degrees in Computer Science and Engineering at the University of Kalyani, India, in 2011 and 2015, respectively. He has co-authored one book and has ten edited books to his credit. He has published over 30 papers in various international journals and conference proceedings, as well as six chapters in edited volumes. His research interests include digital image processing, computational intelligence and hyper spectral imaging. He is a member of ACM, a life member of the Computer Society of India and currently, the regional student coordinator (RSC) of Region II, Computer Society of India.

Dr. Aynur Unal educated at Stanford University (class of '73) comes from a strong engineering design and manufacturing tradition, majored in Structural Mechanics-Mechanical Engineering-Applied Mechanics and Computational Mathematics from Stanford University. She has taught at Stanford University till mid 80's and established the Acoustics Institute in conjunction with NASA-AMES research fellowships funds. Her work on "New Transform Domains for the Onset of Failures" received the most innovative research awards. Most recently, she is bringing in the social responsibility dimension into her incubation and innovation activities by getting involved in social entrepreneurial projects on street children and ageing care. She is also the strategic adviser for Institutional Development to IIT Guwahati. She is a preacher of Open Innovation and she always encourages students to innovate and helps them with her support.

Dr. Santanu Kumar Sen, B.E. (CSE), M.Tech. (CSE), Ph.D. (Engg.), MBA (IS), C.Eng. (I), FIET (UK), FIE (I), FIETE (I), SMIEEE (USA), SMCSI, LMISTE, MACM (USA), has 25 years of experience in the field of Computer Science and Engineering in which 8 years in Industry and 17 years in Engineering Academia including abroad, experience in system level programming/network programming. C/AL/TCP/IP/UDP/IP/TFTP/ICMP/BSD Sockets, Shell Scripting/awk/Perl. Knowledge of Unix Internals and Kernel level programming, SIGNAL/IPC/RPC/File System Management/vnode/vfs interface, R/W Locks/TPI/TLI management, DDI/DKI interface, STREAMS subsystem, Firewall Programming, Ergonomics/Unix and Linux System Administration/Network Administration, Ph.D. (Engg.) under Computer Science and Engineering from Jadavpur University in 2008, Master in Business Administration (MBA) in Information Systems from Sikkim Manipal University in 2011, securing 1st Class (Grade "Excellent", 81% of marks in aggregate) M.Tech. in Computer Science and Engineering securing 1st Class from CMJ University in 2012, Bachelor of Engineering (B.E.) in Computer Science and Engineering from Regional Engineering College (Now National Institute of Technology), Silchar, 1994, securing 1st Class (66.2%), Chartered Engineering (C.Eng.) from Institute of Engineering in 2010, 10+2 std—1st Division (62%) in 1989, 10 std—1st Division (64.3%) in 1987, Principal, Guru Nanak Institute of Technology Sept 2012—till date, Budge Budge Institute of Technology (A Degree Engineering College) 1 August 2011–31 August 2012. He is the head of the Department of Computer Science and Engineering and Department of Information Technology, Guru Nanak Institute of Technology, January 2005–July 2011; a senior lecturer, Asansol Engineering College, Asansol (A Degree Engineering College), July 2003–December 2004; a software consultant, Vanguard Technologies Limited, UK, November 2002–June 2003,; a software engineer, JBAS System Inc., California, USA, December 2000–October 2002; Engineer CMC Limited, December 1997–December 2000; a lecturer in the Regional Engineering College—Silchar, February 1997 to November 1997; and a system analyst, NetCom India, January 1995–January 1997. He published more than 70 papers in various international and national journals and organized various workshops.

Chapter 1
Text-to-Image Classification Using AttnGAN with DenseNet Architecture

Anunshiya Pascal Cruz and Jitendra Jaiswal

Abstract In this attentional generative adversarial networks (AttnGAN) for text-to-image conversion, we have used a CUB dataset with 12,000 images of 200 different birds with 10 captions for each image (12,000 * 10 captions). In random distribution splitting, we divided the data into training and testing sets with 49.2% of data and 50.8% of data, respectively. This method is able to synthesize fine-detailed images by the use of a global attention that gives more attention to the words in the textual descriptions. Also we have the deep attentional multimodal similarity model (DAMSM) that calculates the matching loss in the generator. Though this work produced images of high quality, there was some loss while training the system and it takes enough time for training. This paper proposes the DenseNet architecture with AttnGAN in order to reduce the loss and training time thereby synthesizing images with more distinct features. This technique was able to reduce the loss by 1.62% and could retrieve faster results by 768 s per iteration than the existing CNN architecture.

Keywords GAN · Attn: GAN · Global attention · DAMSM · Dense net architecture

1 Introduction

Text-to-image conversion is a fundamental problem in many applications. The synthesis of images from the given textual descriptions was done using many techniques like stack GAN, stack++ GAN, LR-GAN, FC-GAN, C-GAN, mirrored GAN, etc. [1–7]. But each technique had its own limitations though they were able to produce high-resolution images. Another technique to synthesize image from the

A. P. Cruz (✉) · J. Jaiswal
Department of Computer Science and Engineering, Jain University, Bangalore, Karnataka 560078, India

J. Jaiswal
e-mail: Jitendra.jaiswal@jainuniversity.ac.in

J. K. Mandal et al. (eds.), *Proceedings of International Conference on Innovations in Software Architecture and Computational Systems*, Studies in Autonomic, Data-driven and Industrial Computing, https://doi.org/10.1007/978-981-16-4301-9_1

natural language techniques is the attentional GAN [8] whose extended work is being reflected in this approach. This method is able to synthesize fine-detailed images by the use of a global attention [9–11] that gives more attention to the words in the textual descriptions. Also we have the deep attentional multimodal similarity model (DAMSM) that calculates the matching loss in the generator [12]. Though this work produced images of high quality, there was some loss while training the system and it takes enough time for training. This paper proposes the DenseNet architecture with AttnGAN in order to reduce the loss and training time thereby synthesizing images with more distinct features. In earlier model, the inception architecture [5, 13] has been used, but it resulted in loss during training and consumed a lot of time because the distance between output layer and input layer was longer. In DenseNet, each layer [14–16] is connected to each other where the input of one is fed to the other. This technique was able to reduce the loss by 10% and could retrieve faster results. Also we have the DAMSM that calculates the matching loss in the generator. In this, we have used a CUB dataset with 12,000 images of 200 different birds with 10 captions for each image (12,000 * 10 captions). In random distribution splitting, we divided the data into training and testing sets with 49.2% of data and 50.8% of data, respectively.

2 Literature Survey

T. Xu et al., 2017 worked to generate high-resolution images from the descriptions of the input text. It dealt with multi-stage refinement of fine-grained text-to-image generation using the attentional generative adversarial network that can synthesize fine-grained details at different sub-regions of the image by giving attention to the words in the natural language description. They have additionally used the DAMSM to compute the loss while training the generator. It has boosted the inception score by 14.14% on the CUB dataset and 170.25% on the complex COCO dataset [8]. I. J. Goodfellow et al., 2014 have proposed the generative adversarial network (GAN) for solving many problems in image generation. It basically consists of a generative network and a discriminative network and works by playing a min–max game between two players. The generator tries to fool the discriminator by telling that the generated images are fake and the generator fools vice versa by producing multiple images [17].

Though GAN has shown success in various tasks, it still has a problem when it comes to generating high-resolution images, so Zhang et al., 2018 have proposed stacked generative adversarial networks (stacked GAN). The stacked GAN has two stages of processing. The first stage is the text-to-image synthesis where the shape and the color of the image are sketched to give the low-resolution images. The second stage is for the generation of conditional and unconditional tasks where it contains multiple generators and discriminators. The second stage is more stable than the first stage as it relies on approximation of multiple distributions [3]. W. Zellinger et al., 2019 have proposed the mixture density GAN (MD-GAN) to overcome the problem of collapse in GAN. The discriminator forms clusters, and the generator identifies the various modes of data thereby exploiting it which is achieved by the Gaussian density functions. MD-GAN has the ability to work on collapse issue and generating realistic images using seven datasets [2]. Linyan et al., 2020 proposed an attentional concatenation generative adversarial network (AC-GAN) to generate 1024*1024 high-resolution images from the text. They have used a cascade structure and the DAMSM to calculate the training loss. Added to this they have included diversity measure to the discriminator to increase the diversity in the generated samples. The results have shown that the inception scores on the CUB and Oxford-102 datasets have reached 4.48 and 4.16 that improved by 2.75% and 6.42% compared to AttnGAN [1]. Zhang et al., 2017 proposed conditional augmentation with Stacked GAN to generate photo-realistic images. The diversity of the synthesized image and stability in the training is achieved by the conditional GAN. Extensive qualitative and quantitative analysis has showed better results [4]. Denton et al., 2015 introduced a generative parametric model. This is the LAPGAN which is the integration of the conditional GAN. They have used a Laplacian framework to generate the fine images from the course images. A separate convolution V-Net model is trained by GAN which is used at each step of the pyramid. They concluded the work saying that their model produced 40% times real images compared to 10% of images produced in the GAN [5]. Yang et al., 2017 proposed a layered recursive GAN (LR-GAN) that focuses on both structure of the image and contextual description. It is an unsupervised learning method that generates appearance, pose and shape of the foreground image. The LR-GAN produces better than DC-GAN [6]. Huang et al., 2018 have proposed *face conditional* generative adversarial network (FC-GAN) for generating images of face with high resolutions. This approach produces high-quality images by combining the pixel loss and GAN loss. The generation of these images is not facial expression sensitive. These produced high-quality face images, and it finds a way for many applications like face identification and related ones [7]. Although the deep convolution networks have produced higher accuracy and speed, attainting precise images with detailed features is a problem. In order to resolve this, attempts were made in minimizing the mean squared error, but it unsatisfyingly results in low frequency and high noise in the image. So Ledig et al., 2017 have proposed super-resolution generative adversarial networks (SRGANs) to match this [18].

Ishan et al., 2017 proposed an extension of GAN where multiple discriminators are used instead of one, called the generative multi-adversarial network (GMAN). The role of multiple discriminator ranges from finding the fake image and opposing it to accepting the generated ones. The GMAN produces higher convergence and steady output on various tasks than GAN [9]. Jon Gauthier, 2015 has extended the above work by using conditional GAN (CGAN) which has been used for convolution face generation. This generative model is tuned by varying the conditional information to generate faces with attributes from the random noise. This model was able to control the attributes in the wild dataset by just tuning the conditional information that has given out better results [19]. Y.J. Cao et al., 2018 have discussed the advances of GAN in the field of computer science. Through their survey, they have noted that GAN is efficient in both feature learning and representation than the traditional machine learning algorithms through its adversarial training. They have analyzed the history, development, the mechanism and fundamental network structures of GAN. Several applications of GAN are examined [20]. Initially, deep learning algorithms were used for text-to-image conversion, but the output generated was not efficient. So S. Reed et al., 2016 have used GAN to generate high-resolution images using the discriminator and generator concept. The methods of GAN that were involved in conversion were network architecture for generator and discriminator, GAN-CLS, GAN-INT and inverting the generator for style transfer [21]. Szegedy et al., 2016 and Akshay et al., 2020 have proposed convolution networks that are the core solution for a variety of computer-related tasks and produce high substantial results. However, due to the large-sized model and computational cost, the efficiency and the parameters are low when it comes to mobile vision and big data, and they have discussed inception architecture which can reduce the computational cost and produce better efficiency. However, DenseNet architecture can produce comparatively better results with increased efficiency [22, 23]. Jing yu et al., 2020 have used a dataset that is attached to mouse traces whose narration is grounded to fine-grained user attention. Here they are using a sequential model called TRECS to obtain mask segmentation and to find the object labels linked to the mouse traces. This retrieval-based method is depicted to produce better results than the direct methods [24]. Wah et al., 2011 have created an extension of a bird dataset that has added multiple images to dataset and explains on the dataset [25]. So, B.Gecer et al., 2019 suggested the use of deep convolution neural network (DCNN) along with the GAN in order to produce high-quality images while reconstructing. GAN's generator is used for generation of facial features and 3D morphable model (3DMM) for finding the latent parameters that are supervised by a framework. This has shown to produce better results that retain the reconstructed 3D faces with high frequency [26]. T. salimans et al., 2016 have suggested some new architectures and training features like: (a) feature matching, mini bath discrimination, historical averaging, one-sided label smoothening and virtual batch normalization in convergent GAN training; (b) assessing the quality of the image; and (c) labeling the image quality in semi-supervised learning. Their experiment of semi-supervised classification was on three datasets, namely MNIST, CIFAR-10 and SVHN. The MNIST dataset has generated samples that could not be differentiated by the humans from the original dataset. The CIFAR-10 dataset generated samples with 21.3% error rate [27]. Arjosky et al., 2015 have explained few methods; particularly some techniques are discussed for understanding the principle behind training the GAN. This talks about the problems in training the GAN like its inability, the perfect discrimination theory, saturation, cost function problem, and the suggestion of distributions and soft metrics to solve the gradient issues and instability [28]. Y. Zhan et al., 2018 have proposed a GAN-based semi-supervised learning for classification of hyperspectral images. This method can fully utilize both limited and sufficient unlabeled samples. Firstly, a 3D bilateral filter is used for extracting the spatial features which is collaborated with the spatial information and can be used in subsequent classification. Secondly, GANs are trained for semi-supervised learning which involves updating the features with samples from generator and increasing the output dimensionality of the classifier.

This method has expressed better results with limited labeled samples on AVIRIS and ROSIS datasets than the previously existed model [29]. Paristto et al., 2016 talked about text to image; we need to look on two things—one is generating the correct visual description and other is the attention to semantic description, i.e., if the semantic description matches the generated image or not. This paper proposes a method to seek attention for the semantic description [30].

Diao et al., 2019 have come up with a mirror GAN that preserves the relationship between the visual and semantic description. It consists of three parts: semantic text embedding module (STEM), the global local collaborative attentional module (GLAM) and the semantic text reconstruction and alignment module (STREAM). This method has shown superior results on the CUB and COCO datasets [31]. Anderson et al., 2018 have proposed a combination of combined bottom-up and top-down attention mechanism that helps in understanding the image better. This work helps in calculating the object level and image regions. The bottom-up approach is R-CNN which is faster and concentrates on the image regions and feature vectors, whereas the top-down approach concentrates on the weight of the feature [32]. Kingema et al., 2014 have proposed variation Bayes with learning algorithm that can handle large datasets under different conditions. This is a two-step process: First is the re-parameterization in lower-bound variation for optimizing the gradient method, and they represent the continuous latent variables per data point in the dataset. So, efficient inference could be achieved by a recognition (inference) model using the lower-bound estimator that is proposed in this method and has shown better results [33].

Doersch et al., 2016 described the variation auto-encoders (VAE) that have been used to solve the complications in the distribution of unsupervised learning techniques over the past years. The VAEs were built because of its standard approximation in neural networks that can be used to train the gradient descent. This was used for generating complicated data like written digits, house numbers, CIFAR images, segmentation, etc. [10]. Agnese et al., 2019 and Prajin et al., 2020 have surveyed under GAN in text-to-image conversion by using various techniques such as Stack GAN, Stack GAN + + , AttnGAN, DC-GAN, AC-GAN, TAC-GAN, HDGAN, Text-SeGAn and storyGAN. Again based on this survey, xian et al., 2017 surveyed on synthesis and editing of images using the GAN and the comparing them with various methods [34–36]. For the synthesis of complex images, Anitha et al., 2017 have proposed object-driven attentive GAN. This process involves two steps. One is generation of the salient features of the object relevant to the textual description through object-driven attention. Second, they have used a faster R-CNN for the discriminator so that it checks the relevance between the image and the context [37]. To enable the use of different architecture and minimize the loss function, Zhoa et al., 2016 proposed energy-based GAN (EBGAN) that distributes the energy attributes in the discriminator such a way that the energies are lower near the manifold data and are higher in the other regions [38]. Jose et al., 2016 and Isola et al., 2016 have proposed CNN to achieve high-resolution images and videos of 1080p that feature the map in LR space and sub-pixel convolution for upscaling the images. This method produces better magnitude results of +0.15 dB on images +0.39 dB on videos compared to previous CNN methods [39, 40]. Liang et al., 2017 generated unlabeled samples using the GAN. This consists of two parts. First is the LSRO for the generation of labels for the outliers; second is the CNN for generating the images in the GAN. This method uses three datasets, namely Market-1501, CUHK03 and Duke MTMC-re ID, and has obtained +4.37%, +1.6% and +2.46% improvement than the other methods [41]. Miriam et al., 2017 proposed as the extended work of GAN that generates images with perceptual quality. The generator is designed to optimize the pixel, feature and texture of the generated image against the natural image. The results of this method on CUB and Oxford-102 datasets depict that the perceptual loss functions have improved realism in the synthesized images [42]. The natural climatic conditions such as rain can degrade the quality of the captured images. Zhang et al., 2019 have proposed an approach that focused on de-raining the image. They have achieved this by using the conditional GAN (C-GAN). The adversarial loss in the GAN is used to produce better results. The generator network is constructed using densely connected. Based on this, ID-GAN was developed [14]. Gao et al., 2015 proposed the dense convolution network (DenseNet) which is connected to each layer of the model so that the input from each layer is carried to the next like feed-forward propagation. DenseNet solved the gradient problem, strengthen the feature and reduce the number of parameters increasing the efficiency of the system [13]. Karol et al., 2015 proposed a deep recurrent attentive writer (DRAW) that works by combining the attention mechanism for generation of images and the auto-encoding framework for construction of complex images. This method has proved best result on MNIST dataset [43].

Jun-Yan et al., 2017 proposed a cycle-consistent adversarial network to solve the problem by mapping the input and output images. This is done by pairing the trained images in the training set. The paired set will be considered and unpaired will be left out. We learn the system through mapping and noting the adversarial loss [44]. Nguyen et al., 2016 proposed that by using a gradient latent in the generator space to increase the activation from one to many neurons in an individual classifier to generate these images. He further extended their previous work in this paper by adding an additional latent code that increases quality and diversity of the images. Also they have concatenated the probability of activation method and class to propose the "plug-and-play GAN" (PPGAN) [45]. Alec Radford et al., 2016 have tried to achieve the same success of CNN with supervised and unsupervised in his work. They have developed deep convolution generative adversarial network (DC-GAN) which has architectural limitations thus making it strong. Their work has produced considerable results, but it lacked stability [46]. Tsung-Yi Lin et al., 2014 developed a new dataset for object recognition. They achieved this by gathering complex images with common objects from everyday scenario. Using segmentation, these labels are labeled. This dataset contained images of 91 objects that even 4-year kid could identify. Their developed dataset proved to be more efficient than datasets like PASCAL, ImageNet and SUN [47]. For "pixel-based remote sensing," Xin Pan et al., 2018 proposed a high-resolution image classification technique based on CNN with restricted conditional random field algorithm (CNN-RCRF). This method avoids the distortions while classifying and was successful on two images of remote sensing with good accuracy and less computation time [48]. Peng Zhou et al., 2016 have proposed a combined architecture with bidirectional long-term sentence dependencies (BLSTM) and derive feature by 2D pooling (BLSTM-2D pooling) and 2D convolution (BLSTM-2DCNN). It has shown better performance for 4/6 tasks with high accuracy [49]. Tim Salimans et al., 2018 have presented an optimal transport GAN (OT-GAN) that reduces the metric distanced between the data distribution and generator distribution by combining the energy distance defined in the feature space and eliminates the unbiased gradients [50]. Mike Schuster et al., 1997 have extended a regular RNN to propose a bidirectional recurrent neural network (BRNN) that does not have any limitations to training which takes place in positive and negative time frames. The proposed method gives better performance than other techniques [51]. Ashish Vaswani et al., 2017 have proposed a simple architecture where the transformer is based on attention mechanism alone performing convolution and recurrence [11]. Rintaro et al., 2016 have used the GAN to solve this problem with a framework "Query is GAN" that makes use of the output of the queries to retrieve the scenes. Though the suitable scenes are retrieved, they are not visually clear [52].

Peng Zhou et al., 2016 have proposed a attention-based bidirectional long short-term memory that words by tracing the sentence from front to back and vice versa so equal attention is given to all the words. The evaluation with SemEval-2010 dataset has given better results [53]. Anjie Tian et al., 2019 proposed rdAttn-GAN with multiple generators and discriminators that pay keen attentions to generate diverse images. It also uses optimization techniques to calculate accuracy and similarity. Thus, the system was effective than the previous models and gave comprehensive results on CUB and COCO datasets [54]. Zhe Gan et al., 2017 proposed a semantic compositional network (SCN) in captioning images. They used LSTM to detect probability, and tags are generated. The weight of the matrices is used to calculate the probability of the generated tags [55]. Stanislaw Antol et al. proposed "free-form and open-ended" VAQ that gives answer to the questions in the image. It concentrates on the whole image from the background to the foreground thereby answering to complex scenes [15]. He Zhang et al., 2018 has proposed "Densely Connected Pyramid Dehazing Network (DCPDN)" which learns the atmospheric light in the image and de-hazes it. The atmospheric conditions are fed into the system to make it a physics-based system. Admiring the DenseNet architecture, they densely connected the encoder–decoder that performs multiple pooling preserving the loss [12]. Qingrong Cheng et al., 2019 proposed a deep fine-grained similarity network (DFSN) that consisted of two networks: LSTM and inception V-3 model that is used for extracting the text and the image features. The DFSN gives the similarity matching score word and text features [56]. For text-to-image generation, Tingting Qiao et al., 2019 proposed a Lecia GAN that consists of three phases. The multiple learning phase using TVE, imagining phase using MPA and the creation phase using CAG. It meets the requirements for semantic consistency and visual images [57]. Scott Reed et al., 2016 have proposed NLP models without any pre-training. It has only raw data familiar to human recognition. This shows good performance in retrieval of text-based image [16]. Simon J'egou et al., 2016 have made an extended version of DenseNet where the datasets are trained without any preprocessing [58]. Ming-Yu

Liu et al., 2016 have proposed the coupled GAN (CO-GAN) to work with multiple images and learn joint probability distribution. Like the other methods, this method does not have any dataset for training the images; instead, this works on finding the joint probability of color and depth of the images [59].

3 Background

Human brain has the capacity of processing the text with their imagination to see what it would look like. Similarly, this is a step taken to in-built the imagination of human vision to computer vision that sets a path to the thinking of text-to-image conversion [56] using the description in the natural language. Here, we have proposed the attentional GAN that gives global attention to the captions to give realistic images. We have used DenseNet architecture for better result; DAMSM [12] is trained using the CNN and RNN [51] to calculate the loss and the generator and discriminator to generate the images.

4 Objective

The main reason for this work is to synthesize high-quality images from the given textual captions and to produce better performance than the other models and to reduce the loss and the training time of the proposed GAN than the previous obtained results of other similar works.

5 Methodology

This method proposes the technique to synthesize image from the natural language; attentional GAN [8] whose extended work is being reflected in this approach. This method is able to synthesize fine-detailed images by the use of a global attention [9–11] that gives more attention to the words in the textual descriptions. Also we have the DAMSM [12] that calculates the matching loss in the generator. Though this work produced images of high quality, there was some loss while training the system and it takes enough time for training. This paper proposes the DenseNet architecture with AttnGAN [8, 14, 15] in order to reduce the loss and training time thereby synthesizing images with more distinct features. In earlier model, the inception architecture [5, 13] has been used, but it resulted in loss during training and consumed a lot of time because the distance between output layer and input layer was longer. In DenseNet, each layer is connected to each other where the input of one is fed to the other. This technique was able to reduce the loss by 10% and could retrieve faster results. Also we have the DAMSM that calculates the matching loss in the generator.

6 System Design

The system model of our project forms the plan and rest of the project:

DAMSM: The DAMSM [8, 12] works on two neural networks that map the image region with the sentence sequence in the semantic space to measure the similarity between the text and the image to calculate the loss in generating images.

Text Encoder: We use the bidirectional long short-term memory (LSTM) [1, 44, and 49] as a text encoder to figure the vectors from the word. The feature vector of each word is taken in numbers and dimension. Each word has two hidden states to retrieve the meaning of the word. The global sentence vector is formed by concatenating the last hidden states.

Image Encoder: We use the convolutional neural network to map vectors from the images. The local features of the image are learnt by the middle layers of the CNN, and the global features are learnt by the last layers in the CNN. Our image encoder is built on the DenseNet 201 model [14, 15]. The image is rescaled into pixels to extract the feature matrix. Finally, a perceptron layer is added to convert the image vector into text features.

Global Attention: The attention [9–11] matching score is used to measure the matching relationship image and sentence based on an attention. This attention mechanism concentrates more on the textual content (Fig. 1). Here F_{attn} is the attentional model that performs the upsampling, collects the residuals and reshaping of the output in each stage. The DAMSM loss learns the attention model with semi-supervised algorithm, for matching the entire images with whole sentences. The loss function is calculated by:

$$LossinDAMSM = L_{w1} + L_{w2} + L_{s1} + L_{s2} \tag{1}$$

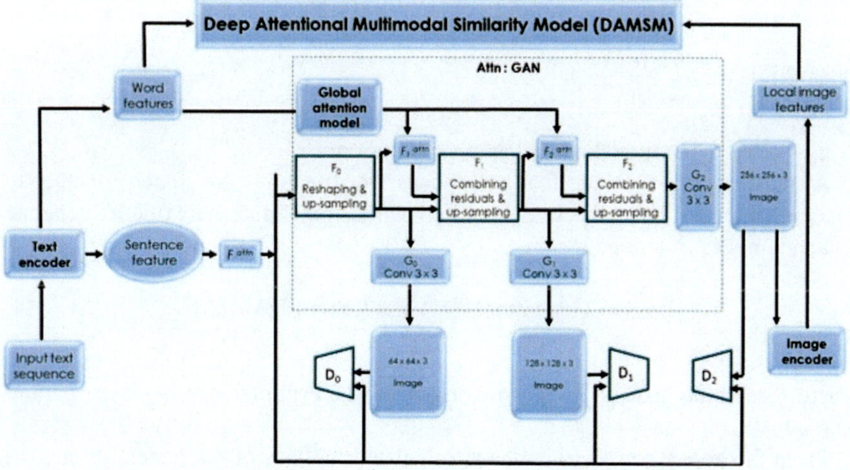

Fig. 1 System architecture

***GAN*:** The GAN [1, 4, 19, 30] consists of a generator network and a discriminator. We have used three generators and three discriminators where the discriminator fools the generator for producing fake images and the generator fools the discriminator by convincing that the generated image is correct. The generators perform the convolution function, and the discriminator checks if the generated images match the text which further verified by the DAMSM.

The function of the generators and their respective hidden states are specifically described as follows:

$$V0 = \boldsymbol{F}0(z, \boldsymbol{Fca}(\overline{\boldsymbol{e}}))$$

$$\boldsymbol{Vi} = \boldsymbol{Fi}(\boldsymbol{Vi} - 1, \boldsymbol{Fiattn}(\boldsymbol{e}, \boldsymbol{Vi} - 1)) fori = 1, 2, ..., m{-}1$$

$$\hat{\boldsymbol{X}}\boldsymbol{i} = \boldsymbol{Gi}(\boldsymbol{Vi}) \tag{2}$$

where

$\overline{e}$—Global sentence vector.

e—Matrix of word vectors.

Z—Noise vector.

F_{ca}—Conditioning augmentation.

V—Sub-region of the image.

V is a feature vector for each column in the image sub-regions. The feature vector is V_{j} for the jth sub-region, and V_{i} for the ith sub-region which is calculated by

$$\boldsymbol{Sj} = \sum_{\boldsymbol{i}=0}^{\boldsymbol{Tx}-1} \boldsymbol{hi}, \boldsymbol{j} \tag{3}$$

where $hj, i = \frac{exp(C_{i,j})}{\sum_{i=0}^{Tx-1} exp(C_{i,j})'}$.

$C'_{j,i} = V_j^T\, e'_i$.

$\mathrm{h_{j,I}}$—Weight the model attends in ith sub-region.

At the next stage, image features and its word features are joined together to generate images. The objective function of the generative network to generate realistic images is defined as

$$\text{Loss} = \text{Generator loss} + \lambda * \text{Loss of DAMSM} \tag{4}$$

where Generator loss $= \sum_{i=0}^{m-1}$(loss of geneator in i subregion); λ—Hyperparameter.

From the above equation, we have calculated the loss of the generator and the discriminator which was combined and used to calculate the total loss of the network.

This loss is the hyperparameter value combined with the loss of the DAMSM. The loss in the discriminator and the generator is the combined error in the dataset. We have calculated the error rate in the CUB dataset during the matching of the word vectors with the image vectors.

Data: In this, we have used a CUB dataset [42] with 12,000 images of 200 different birds with 10 captions for each image (12,000 * 10 captions). In random distribution splitting, we divided the data into training and testing sets with 49.2% of data and 50.8% of data, respectively.

DenseNet Architecture: DenseNet [14, 16] is the modification of CNN architecture where the layers are densely connected. In CNN, the input image is directly passed into a straightforward network layers to get the output. Except the first convolution layer, the output passes into each layer of the network to generate the output image. In DenseNet architecture, each layer in the network is connected to one another layer (all the layers) in the network. The output of each layer is mapped to the input of the next layer and consecutively to the subsequent layers. These nets are more effective because the output is a concatenation of the feature mapping in the previous layers.

7 Analysis and Implementation

Initially as explained in the methodology, we performed feature matching between the semantic text and the image feature so that there is relevance between the both. Feature matching is fixing an objective to the generator in GAN which stops over-training of the discriminator thereby minimizing the output. Thus, a worthy matching takes place. Here are the observations of the feature matching.

Observe the above Fig. 2, which shows the feature matching which the similarity between two datasets by using its distance. Here we can see the feature matching between the CUB dataset which is the target source and the textual description which the name source by using the distance between them. The feature matching completely analyzes the name source and target source and finds a pattern to depict the similarity between them. The similarity may be quality or complexity depending upon the dataset. In our case, it is the similarity between the vectors or weights. Higher vector shows the similarity or closeness between the datasets and uncertainty results in error. This is generally performed to train the generator and discriminator (Figs. 3 and 4).

Then we trained the DAMSM for calculating the loss in order to match the features between the word vector and the image matrix. We obtained the loss of the DAMSM for 200 epochs while training the image encoder and text encoder per batch. We also reduced the execution time and loss in training the images using the inception architecture by replacing it with a DenseNet architecture. The loss of the model using the inception architecture was 0.371, and now using DenseNet is 0.365 for 150 epochs. We have reduced the loss from 0.371 per image to 0.365 per image that would be a total loss of 2222 to 2186 which is the combined loss of the dataset. The cross-entropy

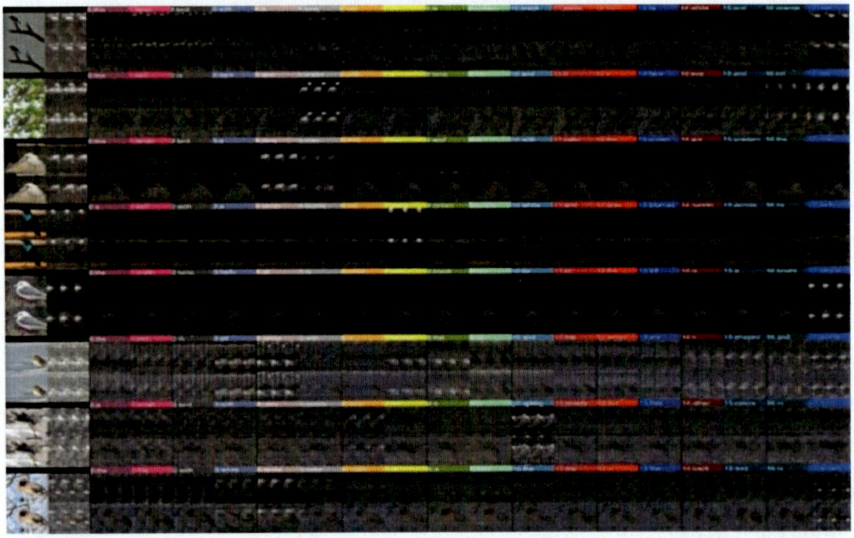

Fig. 2 Feature matching

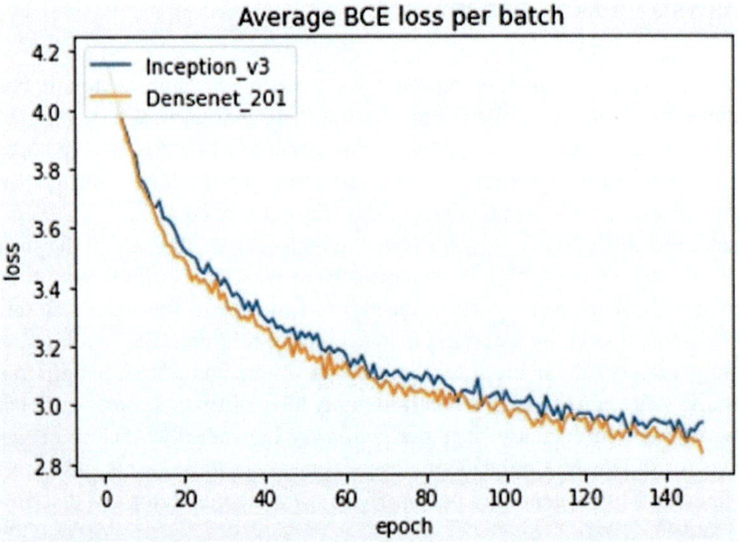

Fig. 3 DAMSM CNN encoder

loss is approximately around 2200 for 749 * 8 images (749 batches of 8 images). From the graph, we can understand that the DenseNet line converges at 150 epochs, but in case of inception it might take 200 epochs to show the convergence which means that the DenseNet architecture is more accurate and faster. The difference

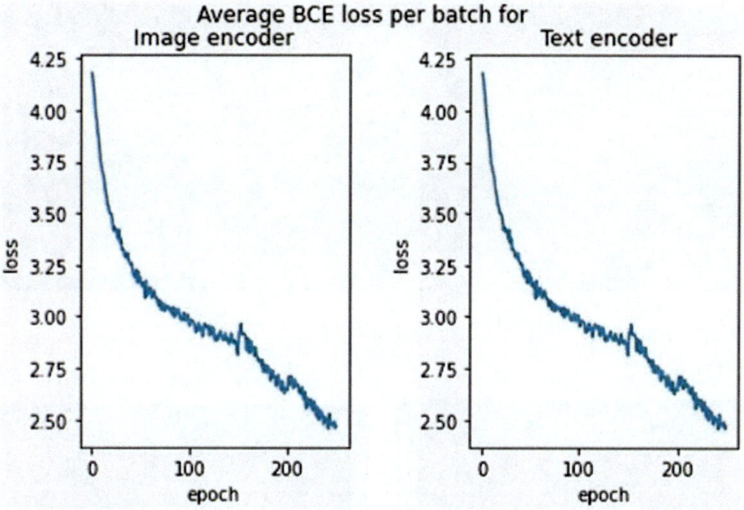

Fig. 4 DAMSM loss

between them is 2.923, 2.811. At 150 epochs, the loss of our model using dense net was 1.62% less than that of inception. The average time taken for one epoch of inception is 3172.14, and for DenseNet is 2404.13. Hence, our model proves to be faster.

8 Results

In this, you can see the image generated from the textual descriptions using the attentional GAN with DenseNet architecture. These are the outcome of the images from their trained weights for the given natural language description. Then we have generated the visual description of the image with respect to its semantic description.

To understand the working of AttnGAN, we can see the above output. Observe Fig. 5; in the first step, the AttnGAN (G0) just draws shape and color of object, but the generated images are of low resolution, and they do not match with the word description as only the vectors in the global sentence are considered. For example, the results in the first stage are corrected in the next stages, namely G1 and G2 based on the textual descriptions like beak and eyes of bird. Thus, the generated images are of higher resolution. For the sub-regions of the images, G1 and G2 are taken directly from the image in the previous state. The attention is distributed to each word and drawn as black in the mapping. For the other regions with more meaningful description, the image drawn is bright with solid colors as shown in the figure. The F_{attn} models help in correcting the images to make in more visible and real by attending the words in the textual description. Therefore, the semantic description

Fig. 5 Image generated by Attn: GAN on CUB dataset

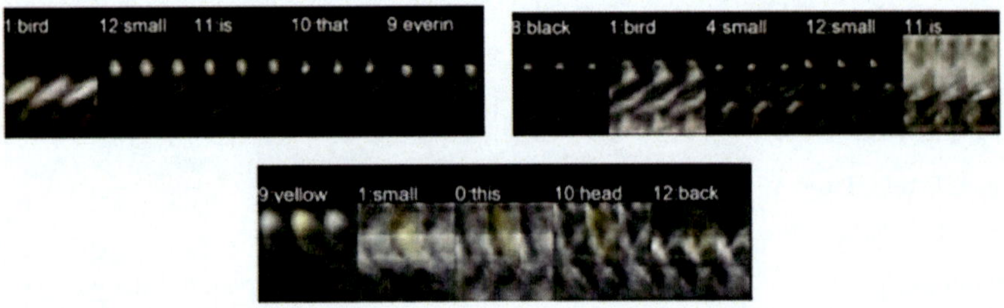

Fig. 6 Trained weights of the images

matches with the output image generated. The $F_{attn\ 2}$ models give attention to the words omitted by the $F_{attn\ 1}$ models. Therefore, the final output images generated are of high resolution.

The trained weight of the images during the training phase is as shown in image 6. In simple words, it is like detecting the objects from the keywords, i.e., the textual descriptions. We have already mentioned that we have a text encoder and image encoder in earlier sections. This takes the text and image, respectively, and converts them into vectors or weights. These vectors are compared, i.e., the word vectors are compared to image vectors to generate similar images. Now let us observe the above Fig. 6; this image shows the matching of these vectors in each stage of the process during the training in order to generate the most relevant images with high resolution.

9 Conclusion

We propose attentional generative adversarial networks (AttnGAN) with DenseNet architecture for text-to-image conversion. This method reduces the training time and loss thereby producing high-resolution images with more distinct features. In this, we have used a CUB dataset with 12,000 images of 200 different birds with 10 captions for each image (12,000 * 10 captions). In random distribution splitting, we divided the data into training and testing sets with 49.2% of data and 50.8% of data, respectively. This method is able to synthesize fine-detailed images by the use of a global attention that gives more attention to the words in the textual descriptions. Also we have the DAMSM that calculates the matching loss in the generator. This technique was able to reduce the loss by 1.62% and could retrieve faster results by 768 s per epoch than inception.

10 Recommendations

This work was able to be carried out for only around 150 epochs due to server and GPU requirements that did not satisfactorily produced results. The number of epochs can be increased in training the system so that the system generates clarified images with defined structures. And further reducing the loss of the DAMSM was challenging and looking forth into other techniques that could be added to update this model.

References

1. Li L, Sun Y, Hu F (2020) Text to realistic image generation with attentional concatenation generative adversarial networks. Hindawi
2. Eghbal-zadeh H, Zellinger W, Widmer G (2019) Mixture density generative adversarial networks. In Proceedings of the Advanced Computer Vision and Pattern Recognition (CVPR), pp 5820–5829
3. Zhang H, Xu T, Li H et al (2018) Stack GAN++: realistic image synthesis with stacked generative adversarial networks. IEEE Trans Pattern Anal Mach Intell 41(8):1947–1962
4. Zhang H, Xu T, Li H et al (2017) Stack GAN: Text to photorealistic image synthesis with stacked generative adversarial networks. In Proceedings of the Advanced International Conference on Computer Vision (ICCV), pp 5907–5915
5. Denton EL, Chintala S, Szlam A, Fergus R (2015) Deep generative image models using a Laplacian pyramid of adversarial networks. NIPS
6. Yang A, Dhruv D (2017) LR-GAN: layered recursive generative adversarial networks for image generation. ICLR
7. Chen H, Lin W (2018) High-quality face image generated with conditional boundary equilibrium generative adversarial networks. Elsevier
8. Xu T, Zhang P, Huang Q, et al (2017) AttnGAN: fine-grained text to image generation with attentional generative adversarial networks. In Proceedings of the Advanced Computer Vision and Pattern Recognition (CVPR), pp 1316–1324
9. Durugkar I, Gemp I, Mahadevan S (2017) Generative multi-adversarial networks. International Conference on Learning Representations
10. Doersch C (2016) Tutorial on variational Auto-encoders. In Proceedings of Statistics and Machine Learning (arXiv), pp 1606–5908
11. Vaswani A, Shazeer N, Parmar N, Uszkoreit L, Jones AN, Kaiser GL, Polosukhin I (2017) Attention is all you need. arXiv:1706.03762
12. Zhang H, Patel V (2018) Densely connected pyramid de-hazing network. CVPR
13. Zhuang H, Laurens L (2015) Densely connected convolutional networks. IEEE

14. Zhang H, Sindagi V, Patel VM (2019) Image de-raining using a conditional generative adversarial network. In IEEE Conference
15. Agrawal A, Lu J, Antol S et al (2017) VQA: visual question answering. IJCV 123(1):4–31
16. Reed S, Akata Z, Schiele B (2016) Learning deep representations of fine-grained visual descriptions. CVPR
17. Goodfellow IJ, Pouget J, Mirza M, et al (2014) Generative adversarial nets. NIPS
18. Ledig C, Theis L, Huszar F, Caballero J et al (2017) Photo-realistic single image super-resolution using a generative adversarial network. CVPR
19. Gauthier J (2015) Conditional generative adversarial nets for Convolutional face generation. IEEE
20. Cao Y, Jia L, Chen Y (2018) Recent advances of generative adversarial networks in computer vision. IEEE 7:14985–15006
21. Reed S, Akata Z, Yan X, Logeswaran L, Schiele B, Lee H (2016) Generative adversarial text to image synthesis. In Proc. Int. Conf. Mach. Learn (ICML), pp 1060–1069
22. Szegedy C, Vanhoucke V, Ioffe S, Shlens J, Wojna Z (2016) Rethinking the inception architecture for computer vision. CVPR
23. Kapoor A, Shah R, Bhuva R, Pandit T (2020) "Understanding inception network architecture For image classification
24. Yu J, Lee J et al (2020) Text-to-image generation grounded by fine-grained user attention. Google Research
25. Wah C, Branson S, Welinder P, Perona P, Belongie S (2011) The Caltech-UCSD birds-200–2011 dataset. Technical Report CNS-TR-2011–001, California Institute of Technology
26. Gecer B, Ploumpis S, Kotsia I, Zafeiriou S (2019) GANFIT: generative adversarial network fitting for high fidelity 3D face reconstruction. In Proceedings of the Advanced Computer Vision and Pattern Recognition (CVPR), pp 1155–1164
27. Salimans T, Goodfellow IJ, Zaremba W, Cheung V, Radford A (2016) Improved techniques for training GAN's. NIPS
28. Arjovsky M et al (2017) Towards principled methods for training Generative Adversarial Networks. In Proceedings of ICLR
29. Zhan Y, Hu D, Wang Y, Yu X et al (2018) Semi-supervised hyper-spectral image classification based on generative adversarial networks. IEEE Geosci Remote Sens Lett 15(2):212–216
30. Mansimov E, Parisotto E, Ba LJ et al (2016) Generating images from captions with attention. ICLR
31. Qiao Z, Xu J, Tao (2019) Mirror GAN: learning text-to-image generation by re-description. In IEEE Conference on Computer Vision and Pattern Recognition (CVPR), pp 1505–1514
32. Anderson P, He XD, Buehler C et al (2018) Bottom-up and top-down attention for image captioning and visual question answering. In Proceedings of the Advanced International Conference on Computer Vision (ICCV), pp 6077–6086
33. Kingma DP, Welling M (2014) Auto-encoding variational Bayes. ICLR
34. Agnese HT (2019) A survey and taxonomy of adversarial neural networks for text-to-image synthesis. Wiley
35. Wu X, Xu K, Hall P (2017) A survey of image synthesis and editing with generative adversarial networks. ISSNll1007–0214ll09/15llpp660–674 Volume 22
36. Jain P, Jayaswal T (2020) Generative adversarial training and its utilization for text to image generation: a survey and analysis. 7(8). ISSN-2394–5125
37. Li W, Zhang P, Zhang L, Huang Q, Gao L (2019) Object-driven text-to-image synthesis via adversarial training. In The IEEE conference on computer vision and pattern recognition (CVPR), pp 12174–12182
38. Zhao J, Mathieu M, Lecun Y (2016) Energy-based generative adversarial network. In Proceedings of ICLR
39. Shi W, Caballero J, Huszar F, Totz J (2016) Real-time single image and video super-resolution using an efficient sub-pixel convolutional neural network. In IEEE Xplore open access
40. Isola Z, Zhou E (2016) Image-to-image translation with conditional adversarial networks. In IEEE

41. Zheng Z, Zheng L, Yang Y (2017) Unlabeled samples generated by GAN improve the person re-identification baseline in vitro. In IEEE Xplore open access
42. Cha M, Gwon Y, Kung HT (2017) Adversarial nets with perceptual losses for text-to-image synthesis. In IEEE Conference
43. Gregor K, Danihelka I, Graves A (2015) DRAW: A recurrent neural network for image generation. In Proceedings of the 32nd International Conference on Machine Learning, JMLR: W&CP, volume 37
44. Zhu J, Park T, Phillip A (2017) Unpaired image-to-image translation using cycle-consistent adversarial networks. ICCV
45. Nguyen A, Yosinski J, Bengio Y, Dosovitskiy A, Clune J (2017) Plug & play generative networks: Conditional iterative generation of images in latent space. CVPR
46. Radford L, Chintala S (2016) Unsupervised representation learning with deep Convolutional generative adversarial networks. ICLR
47. Lin Y, Maire M, Belongie S, et al (2014) Microsoft coco: Common objects in context. In ECCV
48. Pan X, Zhao J (2018) High-resolution remote sensing image classification method based on convolutional neural network and restricted conditional random field. IJRS
49. Peng Z, Suncong et al (2019) Text classification improved by integrating bidirectional LSTM with two-dimensional max pooling. CSCL
50. Salimans T, Zhang H, Radford A, Metaxas D, et al (2018) Improving GAN using optimal transport. ICLR
51. Schuster M et al (1997) Bidirectional recurrent neural networks. IEEE Trans. Signal Processing 45(11):2673–2681
52. Rintaro R, Takahiro MH (2016) Query is GAN: Scene retrieval with attentional text-to-image generative adversarial network. MIC/SCOPE, vol. 4
53. Peng W, et al (2016) Attention-based bidirectional long short-term memory networks for relation classification. In Proceedings of the 54th Annual Meeting of the Association for Computational Linguistics
54. Anjie T, Lu L (2019) Attentional Generative Adversarial Networks with Representativeness and Diversity for Generating Text to Realistic Image. In IEEE
55. Gan Z, Gan C, He X, Pu Y, Tran K, Gao J, Carin L, Deng L (2017) Semantic compositional networks for visual captioning. CVPR
56. Cheng Q, Gu X (2019)Deep attentional fine-grained similarity network with adversarial learning for cross-modal retrieval. Springer
57. Tingting J (2019) Learn, Imagine and Create: Text-to-Image Generation from Prior Knowledge. NIPS
58. Simon M, et al (2016) The one hundred layers tiramisu: fully convolutional dense nets for semantic segmentation. IEEE
59. Ming O (2016) Coupled generative adversarial networks. NIPS

Chapter 2
Cyclone Detection and Forecasting Using Deep Neural Networks Through Satellite Data

Shweta Kumawat and Jitendra Jaiswal

Abstract Satellite imagery provides the initial data information in cyclone detection and forecasting. To mitigate the damages caused by cyclones, we have trained data augmentation and interpolation techniques for enhancing the time-related resolution and diversification of characters in a specific dataset. The algorithm needs classical techniques during pre-processing steps. Using 14 distinct constraint optimization techniques on three optical flow methods estimations are tested here internally. A Convolutional Neural Network learning model is trained and tested within artificially intensified and classified storm data for cyclone identification and locating the cyclone vortex giving minimum of 90% accuracy. The work analyzes two remote sensing data consisting of merged precipitation data from TRMM and QuikSCAT wind satellite data and other satellites for feature extraction. The result and analysis show that the methodology met the objectives of the project.

Keywords Regression · Interpolation · Optical data · Cyclone Intensity · Deep ANN · Convolutional Neural Network

1 Introduction

Most cyclones form over remote ocean areas which are detected on satellite imagery [1]. Earth climate system exhibits variability at different time-related and spatial scales in aspect of cyclone. Remote sensing through satellite data has propagated from meteorology, geological exploration, oceanography and geomorphological surveying on various unconventional zones. Multi-spectral satellite imagery postulates optimized operating time windows, changeover between two continuous imaging task, cloud-coverage effect, etc. The randomly acquired image datasets comprehends the finite time-related resolution which may vary the accuracy of the estimation. One other technique, considered as Dvorak technique which uses infrared

S. Kumawat · J. Jaiswal (✉)
Jain Deemed-to-be University, 562112 Kanakapura, Karnataka, India
e-mail: jitendra.jaiswal@jainuniversity.ac.in

J. K. Mandal et al. (eds.), *Proceedings of International Conference on Innovations in Software Architecture and Computational Systems*, Studies in Autonomic, Data-driven and Industrial Computing, https://doi.org/10.1007/978-981-16-4301-9_2

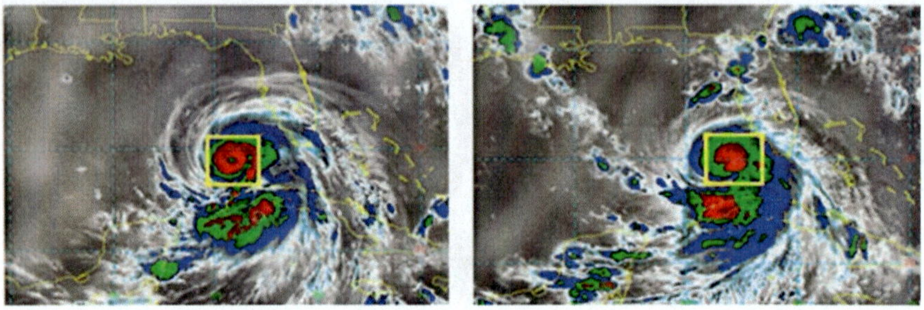

Fig. 1 Satellite imagery

images from geostationary satellites [2]. It evaluates the infrared radiation temperature difference and fetches the value in the eye of cyclone, subtracts the lowest value at 55 km circumference centered at eye, then assigns T-numbers, and a comparison with in situ data shows a good correlation between T-numbers and observed wind speed. But this Dvorak technique is complicated to apply on tropical cyclones because a huge cloud anvil shrouds the center of circulation, so it is impossible to find this center from satellite imagery [3–5]. So, from the perspective of a satellite in space, more intense cyclones have a higher IR 'blackbody' temperature in the eye, and a lower one over the eyewall (Fig. 1).

So expert individuals have scrutinized and analyzed missing data such as a cluster of images portrayed in the storm, a trackway of clouds, vortex site. Hence, the accuracy of analysis relies on time-related resolution of timeframes. The standard image processing techniques are useful but the perception can be replicated using artificial neural network depends on image processing algorithm. In recent surveys, the superiority of deep learning algorithms is used at image processing algorithm in hybrid zones. It is estimated that consequence can be scaled using spatialtime-related interpolation technique via hybridizing the employed forecasted models using deep neural network.

Deep ANN along with stacked denoising autoencoders is estimated better in forecasting air temperature of cyclone in comparison with ANN. While image preprocessing, a deep neural network consist of de-convolution network and sparse denoising autoencoder are established. Thus, are used for high frequency with extraction of low-level features like contours, edges, and shape, size, color, boundary of objects and rotation angle of the cyclone. Accordingly, training the algorithm postulates large quantity of images including wide set of features classified under observation. In case of non-availability of a huge dataset, a technique of artificially increasing the data using interpolation techniques for developing missing time-frames and enhance the attribute-based data [6, 7].

Interpolation: This DL technique amplifies the detailed description of features and characters for the Deep Learning algorithm. It is used in enhancing the cyclone image quality. Other technique consists of optical flow-based time-related interpolation through backward propagation. Optical flow method is the preferential prospective across cloudy atmosphere to perceive interpolation-based characters for image processing. Evaluating optical flow is done via brightness constancy constraint equation. Observed image velocity via optical flow could be utilized for supervised area investigation to an unsupervised dynamic area investigation.

2 Methodology

The deep learning classifies the satellite images having storm and non-storm category. It locates the eyes of the cyclone so that the regression model can predict the cyclone automatically. The training includes generating equations for processing system, constraints, and preferable input. The equations and constraints have to be calculated by standard optimization methods. The present implementation is done observing the case study of cyclone data. Optical flow estimating techniques are computed for error after time-related interpolation. The model calculates the optical flow comprising (FO) fractional order gradients, preferable at Broxs method and Horn and Schunck method and testing subsequently.

This model densifies optical flow and discontinuing sharp outline for better cloud images. To execute interpolation, optical flow inversion vector is needed and analysis is executed while optical flow vector is inverted using optimization techniques. Performance metrics are (1) mean squared error, (2) mean difference error, (3) disagreements of no. of sites, (4) percentage error, (5) sharpness, and (6) peak signal-to-noise ratio.

Next, deep learning data and artificially enhanced data containing cyclone features are used in classification of the cyclonic weather and evaluating the vortex site of the cyclone. Finally, the linear regression method predicts the cyclone track. Satellite data accessible servers in ISRO and NASA are used for training and testing purposes. The implementation is categorized in two major sections one is time-related interpolation algorithms, and other one is deep learning framework. The interpolation part describes mathematical background for calculating the optical flow and algorithm [8–11].

2.1 *Analyzing Optical Flow and Time-Related Interpolation Algorithm with Mathematical Background*

We have analyzed Horn and Schuncks method for estimating the optical flow. The function is defined as follows [12]:

$$E_{\text{data}}(p) = \int [(\nabla I^T P + I_t)^2 + \alpha^2(|\nabla u^2| + |\nabla v^2|)]\mathrm{d}x\mathrm{d}y \quad (1)$$

where ∇I is image intensity gradient. It is a time-related derivative of image intensity, $P = (u, v)T$ as a velocity vector which is optical-flow component, and $\alpha > 0$.

An Interpolation technique used the calculated mean of optical-flow components. The equation of interpolation describes:

$$I(x, T) = (1 - \delta t).I(x - \delta t, u, t_i) + \delta t.I(x - (1 - \delta t).u - 1, t_i + 1) \quad (2)$$

where, I and x are image intensity and position vector, t_i and $t_i + 1$ are two continuous time instances, and optical flow vector $u = (u, v)$, $\delta t = t - t_i$ and normalized time which is $t_i + 1 - t_i = 1$. For calculating $u - 1$, apply the method named Newton Raphson method. From the FO-based methodology, this method for interpolation and optical flow vectors are calculated.

We require various methods for numerical interpolation and the methods for inverse of optical-flow vector are:

Method 1: MoorePenrose pseudoinverse: Matrix $M(p \times q)$, with $p = qn$ or pq, can be dissolved by single value decomposition in single diagonal matrix $\sum$ and two orthogonal matrices U and V so that $M = U \sum VT$. The pseudo-inverse of the matrix is $M+ = U \sum +VT$. This technique calculates the inverse of the optical flow vector and interpolation is performed.

Method 2: Sprinkled Data Interpolation (SDI): Different technique applied to calculate the inverse of optical-flow vector. In case of forward transformation if described, T from originating image space s to targeted image space t, points containing originating space are mentioned as $s = t + T(t)$. Next, the inverse transformation is obtained as

$$I(s) = \frac{\sum_i W(d_i).I(t_i + T)}{\sum_i w(d_i)} \quad (3)$$

where

$$w(d_i) = \begin{cases} \left(\frac{1}{d_i} - \frac{1}{R}\right)^2 & \text{if } d_i \leq R \\ 0 & \text{otherwise} \end{cases} \quad (4)$$

With each interpolation point, $W(d_i)$ is the distance weight function, (d_i) is the distance of interpolated point from the ith data point and search radius is R.

Method 3: Nearest Neighbor (NN) evaluation: NN forward-transformation is applied in the inverse transformation to obtain inverse of the optical-flow vector. If the closest forward transformed point is out of the source voxel, then the mean voxels of forward transformed points are applied.

Method 4: Thin Plate Spline (TPS) interpolation: Contains a set of k points $p_j(x_j, y_j)$ and heights h_j for $j = 1, 2, \ldots, k$, TPS interpolation describes

$$f(x, y) = a_0 + a_x x + a_y y + \sum_{j=1}^{k} c_j U(x - x_j, y - y_j) \quad (5)$$

Where $c_1, c_2, \ldots, c_k$ are constants a_0, a_x, a_y require to find with property of $f(x_j, y_j) = \nabla h_j, \quad j = 1, 2, \ldots, k$.

Method 5: Sigmoid Function Interpolation (SFI): Sigmoid Function is applied for interpolation. We described a univariate logistic curve as:

$$S(x) = \frac{1}{1 + \mathrm{e}^{-\beta x}} \tag{6}$$

where β is steepness of the logistic curve. Hence, the interpolation resulted in 14 different methods in terms of time, iteration, MDE, NSD, PE, PSNR and sharpness: (1) pseudoinverse, (2) pseudoinverse with local convergence, (3) pseudoinverse with global convergence, (4) SDI, (5) SDI with local convergence, (6) SDI with global convergence, (7) NN, (8) NN with local convergence, (9) NN with global convergence, (10) random with local convergence and (11) random with global convergence, (12) TPS interpolation, (13) KR, and (14) SFI applied.

2.2 Deep Learning Frameworks

For classifying cyclone, Keras model has been trained on dataset and google colaboratory is used for training purposes. The convolutional neural network (CNN) models like Xception, SSD and NasNetMobile, are used for pre-processing the dataset. The methodology for detecting and locating the object is to split the image into various segment and supply each segment to model and the label. To eliminate time and cost for computing, you only look once (YOLO) algorithm is preferred for recognizing the circular rotating section of cyclone. The background errors will be eliminated with YOLO as comparison with faster R-CNN. The YOLO divides input image as $N \times N$ grid cell. The specific grid analyzes the object, if the center is in grid cell. Every grid cell forecast B no. of bounding frames with confidence score. These confidence score describes the accuracy. In each bounding frame, five terms x, y, w, h, with confidence score is predicted, where (x, y) is the frame center of the grid cell. The width and height are predicted for the entire image (Fig. 2) [13, 14].

3 Experimental Results

3.1 Dataset

For performing interpolation, we have used satellite image acquired from KALPANA-I and image data is uploaded from India Meteorological Department (IMD). To classify Deep learning, we download the images of cyclones from IMD from the year 2000 till recent period of increased accuracy with coverage. In the technology devel-

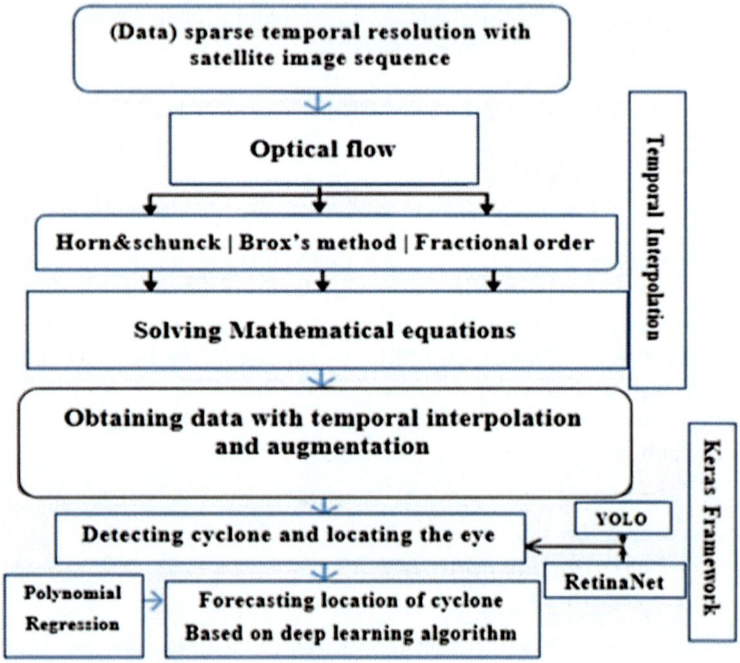

Fig. 2 Background architecture

opment, infrared and mid-infrared, short wavelength infrared, water vapor images and affected cyclones images are clustered in archive. Partial data is chosen for predicting in Table 1.

We have employed the model considering various arbitrary cyclone images uploaded from Meteorological and Oceanographic Satellite Data Archival Center and from Internet. Each image resolution consists of 1024 × 1200 pixel. Implementation of python codes with libraries are used for creating the labels. We investigated this model on cyclone named Ockhi which occurred on 29 Nov 2017 under 3rd Category Hurricane (SSHWS), Very Severe Cyclonic Storm (IMD) and affected areas of Maldives, Sri Lanka, South India with the wind speed of 185 km/h [16–18].

3.2 Observed Results of Interpolation

We have generated various images captured between 05:00:03 and 07:00:03 h. Interpolated images are demonstrated in Fig. 3 (a1) (a13). Left side image in the first range and right-side image in fourth range in Figure (a) and (b) are actual images captured at 05:00:03 and 07:00:03 h. Interpolated images are at almost 8 min of gap from left to right show in the figure. Images at 05:00:03 to 05:24:03 h in first row. The

Table 1 Partial data obtained for interpolation from IMD site [15]

Cyclone name	Year of occurrence	Category	# Of images	Types of images
JAL	2010	Severe cyclonic storm	200	Visual
BOB3	2007	Extremely severe cyclonic storm	130	Infrared
GONU	2007	Super cyclonic storm	115	Infrared
LEHAR	2013	Very severe cyclonic storm	109	Infrared
MADI	2013	Very severe cyclonic storm	150	Visual
BANDU	2010	Cyclonic storm	170	Infrared
AILA	2009	Severe cyclonic storm	57	Visual

image shows from 05:32:03 to 05:56:03 h in second row. Images from 06:04:03to 06:28:03 h in the third row. The images from 06:36:03 to 07:00:03 h in the last row. We Observed that the second image of the first row is similar to image captured at 05:00:03 h and the second last image of the last row is very close to 07:00:03 h captured image.

The analyzed results describe that artificial interpolated images gives a mild variation in cyclones circumference and bit change in diameter at 05:00:03 h into cyclone stated into second image comparatively large diameter with ovate shaped. These are uncaptured dimension, however artificially generates the images, which will be useful in neural network in cyclone classification. Comparatively change of first captured image in the Hausdorff dimension reduced with time series of interpolated image shown in Fig. 3.

3.3 *Training and Results*

We have experimented deep learning for two various significances, one is for classification of an image categorized under cyclone or not carrying any features of existing cyclone and other one is for forecasting the site of cyclone in the future prospective. **Classifying the cyclone and non-cyclonic circumstances**: Tropical cyclones mostly occur in the North Indian Ocean, which strikes the Indian subcontinent mostly in a month of May till mid-December with major loss and severe destruction. The meteorological center of cyclone around North Indian Ocean, IMD regularly tracks the Indian Ocean cyclonic occurrences with its trajectory. As per IMDs identification North Indian Ocean cyclone begins as a depression around the wind speed of 31

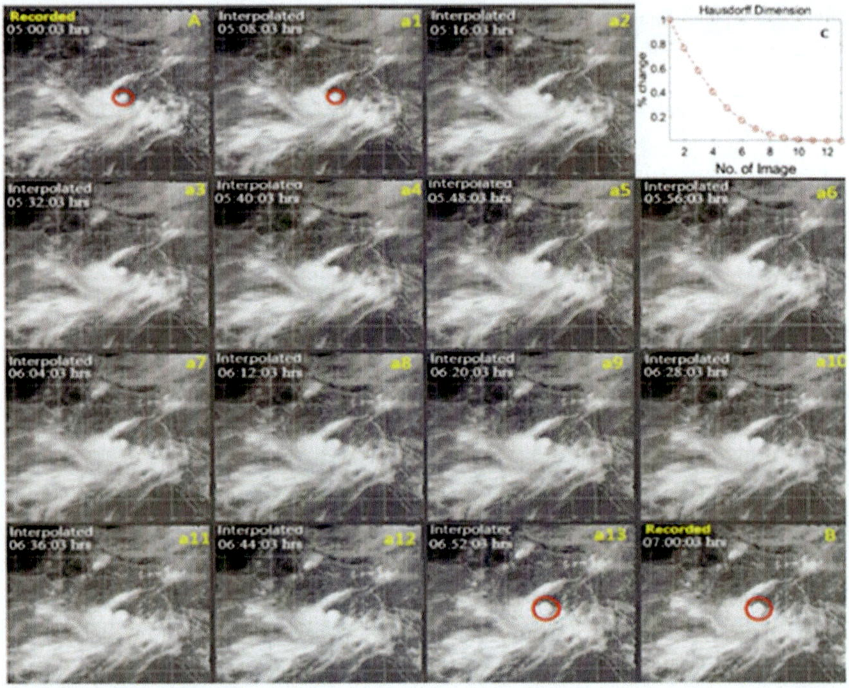

Fig. 3 Interpolation with time series data

50 km/h in the Bay of Bengal and in the Arabian Sea. Depression turns into deep depression when the speed reaches 51–62 km/h and the system into huge moisture content. When the wind speeds redouble across 63–88 km/h with the extensive subsistence, IMD detects it under high cyclonic storm and assigns the category with name. The next part of severe cyclonic storm classifies when the wind speeds summits in an extent of 118,165 km/h and cause heavy destruction. Further investigated observations imply the cyclone is detected into a very severe, extremely severe, and super cyclonic category based on the cyclone intensity. We have used the CNN framework with Keras for the identifying storm images under these five classes provided by IMD [19–21].

The satellite data includes infrared and visual images of eight occured cyclones from IMD are given as input to convolution neural network and the model is employed on Google collaboratory to draw the image features. time-related interpolation at images based on optical flow is applied to augment the data. Before data training, image pre-processing is done. This subtracts the un-complying and noisy characteristics and improves the efficiency of the model. Raw image from the dataset is chopped to eliminate header files with white edges. In Fig. 4 Simultaneously binarizing the image carrying additional modifications which is used on the images [22].

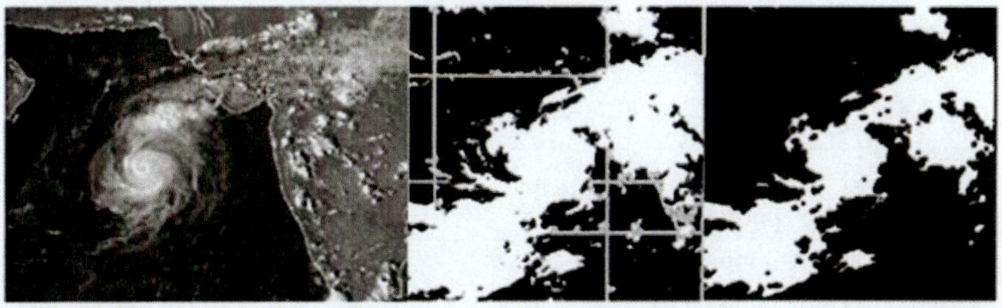

Fig. 4 Left image is satellite image, middle image is binarized and right image is after erosion

Purpose of this point is to remove the unwanted information such as grid lines, geographical boundaries, and landscape. In image binarization, the image is modified in a binary image along with pixel intensities above threshold are modified to one and another to zero. In this algorithm, the output will be again RGB image with the pixels intensities high upon a threshold value maintaining the actual values meanwhile other pixels are modified to lowest minima which will be zero. In result, the vortex and peripheral cloud blots were maintained along with actual pixel values and other features were subtracted. On the further processing image erosion is used (Fig. 4).

3.4 *Model Training and Output*

Around thousand images are uploaded from IMD but data required to train the model, data augmentation is needed. Optical flow-based on time-related interpolation is used on images and ten images were generated between two time-relatedly continuous images. Keras model is employed on the preprocessed dataset. This platform of Google collaboratory was applied for training purpose and it trains the 6910 augmented images with 2960 images verified and unorderly shuffles and divided from the actual database. The verification accuracy is 97% obtained for classifying cyclone and non-cyclonic weather conditions [23, 24].

Table 2 represents sequential CNN model constructed and trained for classifying images of remote sensing data. This model and Xception results true outcomes. For further work, the data can be augmented with different features of wind and temperature and a merged RNN and CNN model can be applied for better classification of the cyclone [17, 25].

4 Detection Based on Remote Satellite Data

We have used remote sensing measurements in some parts of implementation which is QuikSCAT wind satellite measurements, and merged precipitation data belong to TRMM with various satellites.

Table 2 CNN model perspectives

S. No.	Type	Channel of filters	Filter/Pool size	Activation
1	Convolutional	32	3.3	ReLU
2	Convolutional	32	3.3	ReLU
3	Max. pooling	–	2.2	–
4	Convolutional	64	3.3	ReLU
5	Convolutional	64	3.3	ReLU
6	Max. pooling	–	2.2	–
7	Dense	128	–	ReLU
8	Dropout	0.3	-	-
9	Dense	5	–	Softmax

QuikSCAT wind data generates information from the TRMM precipitation data for cyclone site for TRMM detector to image on search at specific region. This part contains two categories of remote sensing data used in our cyclone prediction and tracking implementation. QuikSCAT wind data from polar orbiting satellite and the combined high quality/infrared precipitation data from the TRMM orbiting satellite and another Geostationary Operational Environmental Satellites (GOES) [13].

4.1 QuikSCAT Wind Data and TRMM Satellite Precipitation Data

The QuikSCAT gives high quality ocean wind dataset. QuikSCAT is a polar orbiting satellite consisting 1800 km wide dimension swath over the Earth surface. It givesthe output two times per day which covers over a specified global region. The microwave radar which is SeaWinds instruments on the QuikSCAT satellite calculates the wind speeds and direction under the weather with cloud conditions hover the global oceans. The TRMM satellite contains five remote sensing instruments are Precipitation Radar (PR), TRMM Microwave Imager (TMI), Visible Infrared Scanner (VIRS), Clouds and Earth Radiant Energy Sensor (CERES), and Lightning Imaging Sensor (LIS) [11, 26, 27].

It orbits between 35° north and 35° south of the equator. It measures between 50° north with 50° south of the equator (Fig. 5).

Tracking of cyclone using the QuikSCAT and the TRMM data (Latitude: 0–50 N; Longitude: 30–80 W). Predicted cyclone squared in a frame [14].

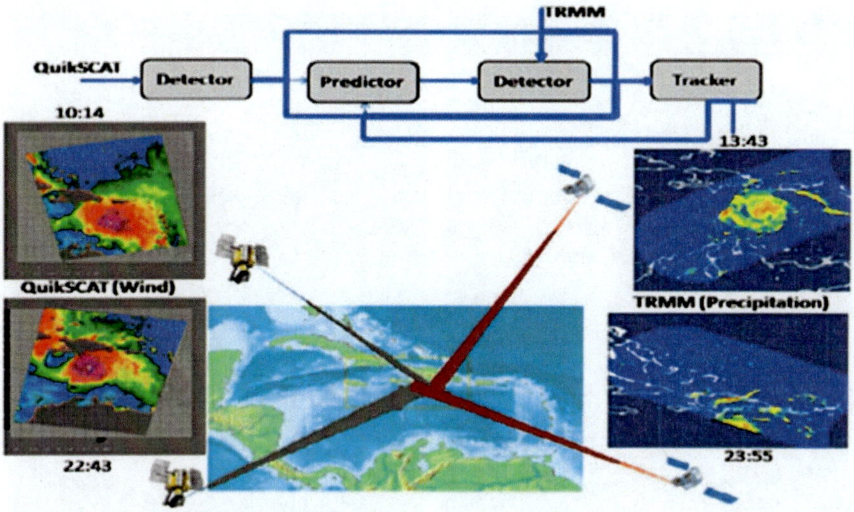

Fig. 5 Measurements of orbits on equators

4.2 *QuikSCAT Feature Extraction*

In cyclone detection and tracking, features which characterize and detects a cyclone are pointed which are useful in extracting feature from QuikSCAT satellite data. We use the QuikSCAT Level 2B data involving ocean wind vector data arranged through complete orbital revolution of the satellite. One satellite whole polar orbiting revolution consists of 101 min. The Level 2B data are clustered by range in wind vector cells (WVC) of squares dimension 25 km or 12.5 km (Fig. 6).

The whole earth circumference needs 1624 WVC rows and spatial resolution of 25 km, around 3248 rows at 12.5 km spatial resolution. The 1800 km swath width amounts to 7225 km WVCs or 144 12.5 km WVCs. Occasionally, the measurements lie outside the swath. Hence, the Level 2B data contains 76 WVCs at 25 km spatial resolution and 152 WVCs at 12.5 km spatial resolution to adapt those instances. There exist twenty-five zones in the data structure for the Level 2B data. We are captivated in evaluating in the latitude, longitude, and the wind speed and direction for the WVCs. The zones are mentioned in Table 3 [22, 28, 29].

Therefore, the Level 2B data is acquired, it requires interpolation on uniform grid surface due to non-uniformity at dimensions of QuikSCAT satellite on spherical surface. The nearest neighbor rule is applied then for pre-processing for wind speed and direction. Consider the wind speed and wind direction at site. One defines the direction to speed ratio (DSR) at when a strong wind across wind circulation is present, the DSR of wind vector cell can be minor. A histogram is generated to calculate the possibility of DSR in a region will have a slant distribution over the smaller value. When weak or no wind with no circulation is found, DSR histogram does not have the slant features [29].

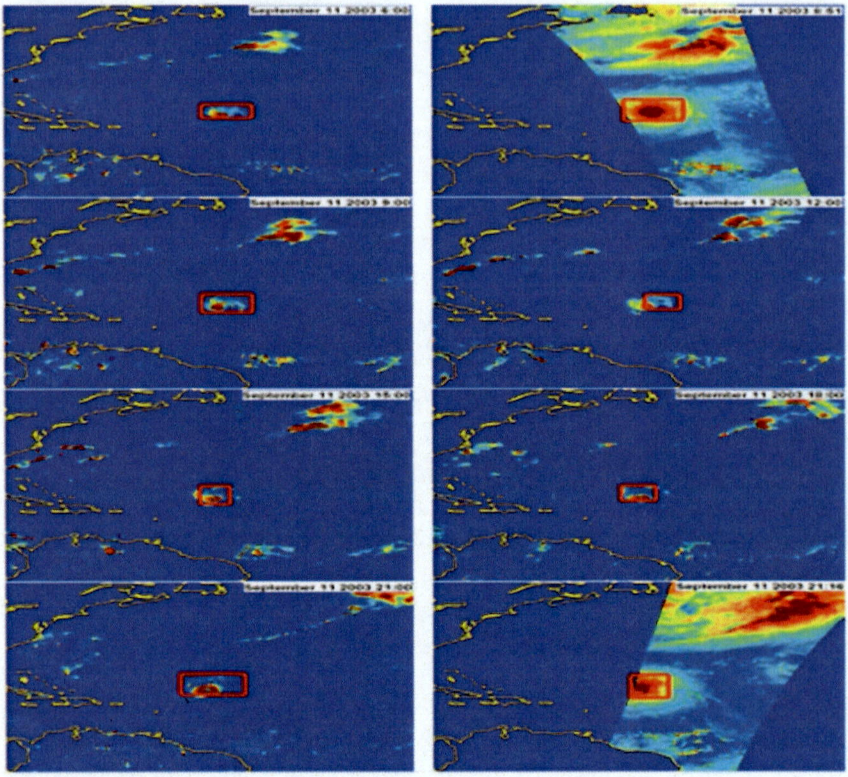

Fig. 6 The level 2B data clustering by range in wind vector cells (WVC)

Therefore, we train QuikSCAT for cyclone detection in every twelve hours and TRMM data for tracking in every three hours based on QuikSCAT features. QuikSCAT data will be received from the current streaming information, and inputs in cyclone detection module to detect probability of cyclones. The cyclone site, using a linear Kalman filter predictor, then applied to predict the regions of cyclone occurrence at the next incoming data stream received. The cyclone is located after applying a threshold to TRMM precipitation measured rate (T6 = 0). When a cyclone is located in the TRMM data, the applied filter measurement upgrades the correction, which is used acquire a forecasted site of the cyclone in next TRMM (or QuikSCAT) observation in next three hours of intervals [11].

5 Tracking and Forecasting Cyclone

Here, we specified a model to track the eye or the center of the cyclone. The objective of model is to analyze the patterns captured in its movement and then predict

Table 3 The zones

Field	Unit	Minimum	Maximum
WVC latitude	Deg	−90.00	90.00
WVC latitude	Deg E	0.00	359.99
Selected speed	m/s	0.00	50.00
Selected direction	Deg from North	0.00	359.99

the path of the cyclone. So we need to physically interpret the bounding frames of cyclone images. We obtained the cyclone images and interpreted it through Matplotlib (python library). We detected hundred images but deep neural network can fit on this small dataset. So to evaluate this, we require to depend on augmentation. Subsequently augmentation, we formed five thousand images for employing and five hundred images for testing. Cyclone images given by the Indian Meteorological Department are used. These satellite images were recorded in half an hour interval, with various spectrum that is visible and infrared. Then used interpolation technique to obtain images. This model contains two distinct phase that are Training and forecasting [7, 24].

The deep learning model Retina Net needs CSV file which includes the following format. path2image, x1, y1, x2, y2, obj, and path2image explains the entire path of the image, and (x1, y1) , (x2, y2) coincides the top-left points with bottom-right points of bounding frame.. As finally it results in the objective to detect a cyclone (Fig. 7).

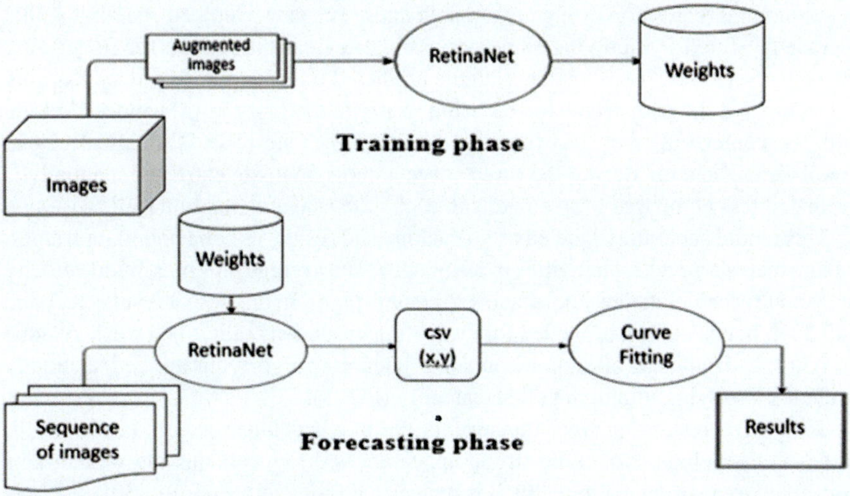

Fig. 7 Training and forecasting phases

Prediction phase: We need impart two CSV files: one for training and other for verification. Training is done at NVIDIA GTX 1080 graphics card including 24-h process time. The weights corresponding to network are saved. It generates the bounding frames over the detected cyclones, as we are concerned about locating the centers of cyclone, we calculate the center coordinates based on top-left and bottom-right points and saved to CSV file. Therefore, other deep learning model that is long short-term memory can be applied to extract the time-related compendium. For employing an LSTM, high-frequency dataset is required which is not available. The data of cyclone named Ockhi obtained by MOSDAC which gives the final evaluation [29].

5.1 Case Study

The case study from 29 November 2017 to 5 December 2017 gives the the presence of a cyclone and found departing from Indian Ocean through Peninsular India next to Arabian sea. The data obtained containing fourty four images for everyday in an interval of thirty minutes. Using interpolation technique, image set is dense around hundred images, 15-min interval of every image. Figure 8 describes (a1) (a8) which shows images of complete day. (b1) includes overlaid images (a1) and (a8). Next (c1) containing zoom area, in a presence of two centers of cyclone, vortex Vc1 and Vc2. It displays the path of the eye in entire day, evaluated through deep learning algorithm and traced the path manually. (d1) shows isolated trajectories in red color bigger circles that is predicted through two methods. The algorithm evaluates coordinates of ninety-five centers, including sixty-one with exclusive coordinates which we applied in polynomial regression fitting next to both ends. The data averaged containing blue speckle are fitted by applying same polynomial regression method for forecasting the path of cyclone. It will match with the predicted track acquired by deep learning algorithm and the path seems to be arising nearly to the first eye (a1) and (d1) along with the center (a8) and (d8) from the final image. One pixel is identical to one km. Average spacing during the fitted curve and original site extracted manually is evaluated less compared to five pixels in each case as per estimation [30].

This model accurately labeled as no cyclone and results as none bounding frames, if there theres no cyclone activities in the images. The average plus maximum velocity of cyclone is calculated by coordinates of centers (eye). In this case it results 30.1 and 140.2 km/h, correlation is needed for evaluating grounded reality. The weak cyclone at right side disappears after nine hours. Deep learning algorithms applied, distinctly exclude classifying it under cyclone category (Fig. 8).

Table 4 Deep learning algorithm evaluations of all cyclones occurred in 2016–17. In few case cyclone containing strong appearances disappears out and weaker gets converged to a strong cyclone. Such instances are tested successfully [30].

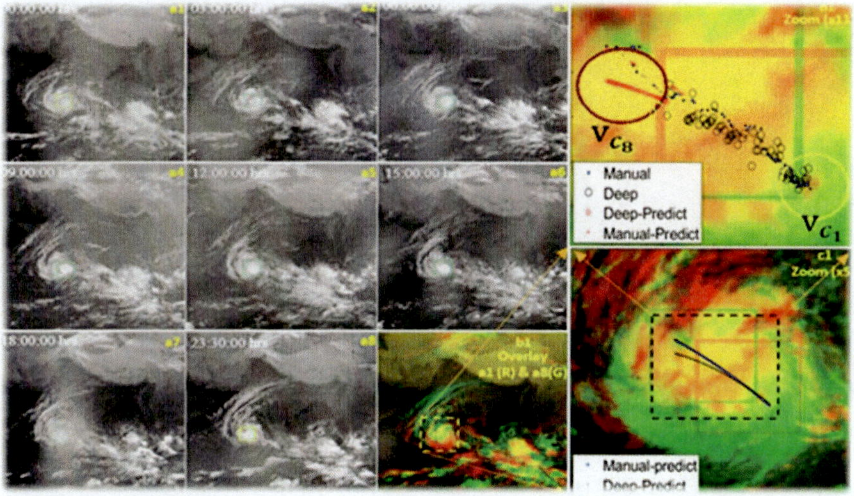

Fig. 8 Cyclone category

Table 4 Deep learning algorithm evaluations of all cyclones occurred in 2016–17

S. No.	Cyclone occurred	Year of occurrence	RMSE (%)
1	ROANU	17 May–21 May 2016	15.51
2	KYANT	25 Oct–27 Oct 2016	6
3	VARDAH	07 Dec–12 Dec 2016	9
4	MORA	27 May–30 May 2017	5
5	OCKHI	29 Nov–7 Dec 2017	10

6 Observations

As per the figure the description of five various cyclones elevated in this time span and data is acquired from the Joint Typhoon Warning Center (JTWC) and shown in Fig. 9 JTWC data tracks are procured from web repositories of NCAR Lab, UCAR containing yellow pinned form in Ockhi case and North Indian Ocean data paths are from Naval Meteorology and Oceanography which contain multiple color cyclone shape, eye coordinates of five cyclones are acquired manually and shown using red dots contains data accommodated to Ockhi in Arabian Sea side in Fig. 10. The other four cyclones are extant in Bay of Bengal. Root-mean-square error is calculated during coordinates acquired through deep learning and estimated manually.

The deep learning algorithm successfully detected the event and cyclone activities in two years of time in some sites which rematches to manually acquired with minor error of 15%. Figure (c) shows according to date color-labelled traced path of coordinates carrying cyclone eye using this algorithm with distinct colors which are green, pink, blue, yellow-red, blue, yellow and green since 29 November to 4

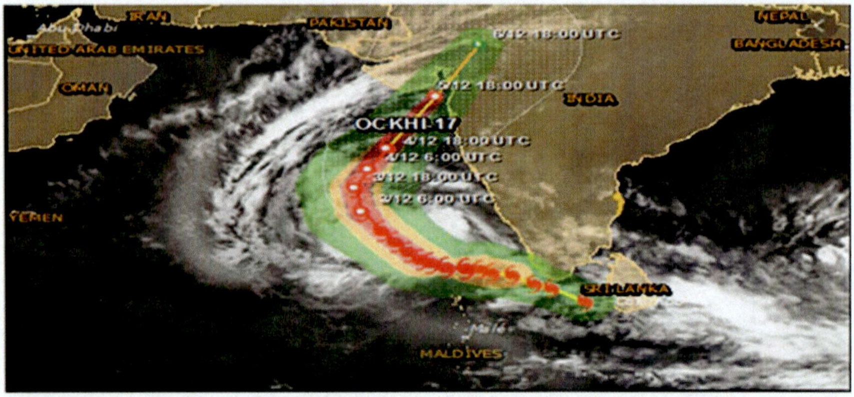

Fig. 9 The representation of five various cyclones elevations

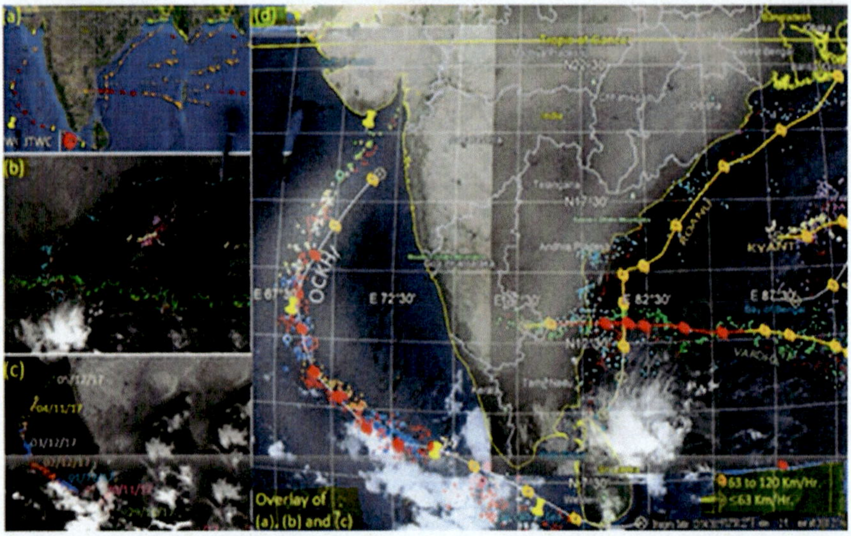

Fig. 10 The Ockhi in Arabian Sea side

December 2017.It is estimated in figure that deep learning evaluation and manual calculations are identic toward each other but JTWC path strays whenever cyclone occurs weak. The slope of the hind section of Mora (cyclone) and Ockhi varies. We have observed that the JTWC data by two resources for Ockhi in which one stated in yellow color dot and other through cyclonic form markers[13, 30].

7 Conclusion

The convolution neural network model effectuates classical models in the detection of remote sensing images. Specifically, the YOLO model is considered for detection and site the cyclone. R-CNN model is preferable for detecting the site of the cyclone. For interpolation, the FO-based approach is quite better than Broxs method for optical flow evaluation. RetinaNet is a better model than LSTM If the high-frequency dataset is not accessible and available. The analysis of deep learning algorithm rectifies the huge dataset by interpolation as well as data augmentation. Tracking and estimating the center of cyclone observed quite accurate using deep learning algorithm in contrast with manual process, if images contain one cyclone. Successfully, deep learning algorithms for detecting and forecasting the storm in a future are completely tested and verified. For identifying the cyclone and non-cyclonic condition the accuracy of 96% and detecting cyclone results in accuracy obtained higher than 85%.

References

1. Kovordanyi R, Roy C (2009) Cyclone track forecasting based on satellite images using artificial neural networks. ISPRS J Photogram Remote Sens (Print) 64(6):513–521. https://doi.org/10.1016/j.isprsjprs.2009.03.002
2. Zhang J, Zhong P, Chen Y, Li S (2014) L (1/2)-regularized deconvolution network for the representation and restoration of optical remote sensing images. IEEE Trans Geosci Remote Sens 52(5):2617–2627. https://doi.org/10.1109/TGRS.2013.2263933
3. Langella G, Basile A, Bonfante A, Terribile F (2010) High-resolution space time rainfall analysis using integrated ANN inference systems. J Hydrol 387(3–4):328–342. https://doi.org/10.1016/j.jhydrol.2010.04.027
4. Li J, Huang X, Gong J (2019) Deep neural network for remote-sensing image interpretation: status and perspectives. Natl Sci Rev 6(6):1082–1086. https://doi.org/10.1093/nsr/nwz058
5. Valizadeh N, Mirzaei M, Allawi MF et al (2017) Artificial intelligence and geo-statistical models for stream-flow forecasting in ungauged stations: state of the art. Nat Hazards **86**, 1377–1392. https://doi.org/10.1007/s11069-017-2740-7
6. Zhao X, Xu T, Fu Y, Chen E, Guo H (2017) Incorporating spatio-time-related smoothness for air quality inference. In: IEEE International Conference on Data Mining (ICDM), New Orleans, LA, pp 1177–1182. https://doi.org/10.1109/ICDM.2017.158
7. Grover A, Kapoor A, Horvitz EJ (2015) A deep hybrid model for weather forecasting. In: Proceedings of the 21th ACM SIGKDD international conference on knowledge discovery and data mining. KDD'15, Sydney, NSW, Australia, pp 379–386. https://doi.org/10.1145/2783258.2783275
8. Zhang L, Zhang L, Du B (2016) Deep learning for remote sensing data: a technical tutorial on the state of the art. IEEE Geosci Remote Sens Mag 4(2):22–40. https://doi.org/10.1109/MGRS.2016.2540798
9. Parker JA, Kenyon RV, Troxel DE (1983) Comparison of Interpolating Methods for Image Resampling. EEE Trans Med Imaging 2(1):31–39. https://doi.org/10.1109/TMI.1983.4307610
10. JTWC track: Naval meteorology and oceanography command. https://www.metoc.navy.mil/jtwc/jtwc.html?northindian-ocean. Last accessed 4 Oct 2020
11. National Center for Atmospheric Research Database: http://hurricanes.ral.ucar.edu/realtime/plots/northindian/2017/io032017/. Last accessed 4 Oct 2020

12. Jan E, Dennis S, Heinz H (2006) Interpolation of time-related image sequences by optical flow based registration. https://doi.org/10.1007/3-540-32137-3_52
13. Samy M, Karthikeyan SK, Durai S, Sheriff R (2018) Ockhi cyclone and its impact in the Kanyakumari district of Southern Tamilnadu, India: an aftermath analysis. Int J Recent Res Aspects 466–469
14. Johnson JM, Khoshgoftaar TM (2019) Survey on deep learning with class imbalance. J Big Data 27:466–469. https://doi.org/10.1186/s40537-019-0192-5
15. Shakya S, Kumar S, Goswami M (2020) Deep learning algorithm for satellite imaging based cyclone detection. IEEE J Sel Top Appl Earth Observ Remote Sens 13:827–839. https://doi.org/10.1109/JSTARS.2020.2970253
16. Fousiya AA, Lone AM (2018) Cyclone Ockhi and its impact over Minicoy Island, Lakshadweep, India. Current Sci. 115(5):819–820
17. Pao TL, Yeh JH (2008) Typhoon locating and reconstruction from the infrared satellite cloud image. J Multimedia 3(2):45–50. https://doi.org/10.4304/jmm.3.2.45-51
18. Yunjie L, Evan R, Prabhat, Joaquin C, Amir K, David L, Kenneth K, Michael W, William C (2016) Application of deep convolutional neural networks for detecting extreme weather in climate datasets. arXiv:1605.01156
19. Jolliffe IT, Stephenson DB (2003) Forecast verification: a practitioners guide in atmospheric science. Wiley, Hoboken
20. Langella G, Basile A, Bonfante A, Terribile F (2010) High-resolution space time rainfall analysis using integrated ANN inference systems. J Hydrol 387(3–4):328–342. https://doi.org/10.1016/j.jhydrol.2010.04.027
21. Liu Q, Wu S, Wang L, Tan T (2016) Predicting the next site: a recurrent model with spatial and time-related contexts. In: AAAI'16: proceedings of the thirtieth AAAI conference on artificial intelligence, Phoenix, Arizona, pp 194–200
22. Redmon J, Divvala S, Girshick R, Farhadi A (2016) You only look once: unified real-time object detection. In: Proceedings of IEEE conference computer vision and pattern recognition. arXiv:1506.02640
23. Panigrahi CR, Sarkar JL, Pati B (2018) Transmission in mobile cloudlet systems with intermittent connectivity in emergency areas. Digital Commun Netw 4(1):69–75
24. Huttenlocher DP, Klanderman GA, Rucklidge WJ (1993) Comparing images using the Hausdorff distance. IEEE Trans Pattern Anal Mach Intell 15(9):850–863. https://doi.org/10.1109/34.232073
25. Panigrahi CR, Sarkar JL, Pati B, Bakshi S (2016) E3M: an energy efficient emergency management system using mobile cloud computing. In: IEEE International conference on advanced networks and telecommunications systems (ANTS), Bangalore, pp 1-6
26. Panigrahi, CR, Tiwary M, Pati B, Das H (2016) Big data and cyber foraging: future scope and challenges, techniques and environments for big data analysis. In: Studies in big data, vol 17. Springer, pp 75–100
27. Jia X, Kuo BC, Crawford MM (2013) Feature mining for hyperspectral image classification. Proc. IEEE 101(3):676–697
28. Panigrahi CR, Sarkar JL, Tiwary M, Pati B, Mohapatra P (2019) DATALET: an approach to manage big volume of data in cyber foraged environment. J Parallel Distributed Comput 131:14–28
29. Melgani F, Bruzzone L (2004) Classification of hyperspectral remote sensing images with support vector machines. IEEE Trans Geosci Remote Sens 42(8):1778–1790
30. Brox T, Bruhn A, Papenberg N, Weickert J (2004) High accuracy optical flow estimation based on a theory for warping. In: Proceedings of European conference on computer vision, vol 4), pp 25–36

Chapter 3
An Improved Differential Evolution Scheme for Multilevel Image Thresholding Aided with Fuzzy Entropy

Rupak Chakraborty, Sourish Mitra, Rafiqul Islam, Nirupam Saha, and Bidyutmala Saha

Abstract Image segmentation problem has been solved by entropy-based thresholding approaches since decades. Among different entropy-based techniques, fuzzy entropy (FE) got more attention for segmenting color images. Unlike grayscale images, color images contain 3-D histogram instead of 1-D histogram. As traditional fuzzy technique generates high time complexity to find multiple thresholds, so recursive approach is preferred. Further optimization algorithm can be embedded with it to reduce the complexity at a lower range. An updated robust nature-inspired evolutionary algorithm has been proposed here, named improved differential evolution (IDE) which is applied to generate the near-optimal thresholding parameters. Performance of IDE has been investigated through comparison with some popular global evolutionary algorithms like conventional DE, beta differential evolution (BDE), cuckoo search (CS), and particle swarm optimization (PSO). Proposed approach is applied on standard color image dataset known as Berkley Segmentation Dataset (BSDS300), and the outcomes suggest best near-optimal fuzzy thresholds with speedy convergence. The quantitative measurements of the technique have been evaluated by objective function's values and standard deviation, whereas qualitative measures are carried out with popular three metrics, namely peak signal-to-noise ratio (PSNR), structural similarity index measurement (SSIM), and feature similarity index measurement (FSIM), to show efficacy of the algorithm over existing approaches.

Keywords Color image · Multilevel thresholding · Fuzzy entropy · Differential evolution

R. Chakraborty (✉) · S. Mitra · R. Islam · N. Saha · B. Saha
Department of CSE, Guru Nanak Institute of Technology, Kolkata, India
e-mail: rupak.chakraborty@gnit.ac.in

J. K. Mandal et al. (eds.), *Proceedings of International Conference on Innovations in Software Architecture and Computational Systems*, Studies in Autonomic, Data-driven and Industrial Computing, https://doi.org/10.1007/978-981-16-4301-9_3

1 Introduction

Segmentation of images appears as a preliminary step to analyze or interpret images mostly applicable in computer vision-oriented applications to extract objects from its backgrounds. In the segmentation approach, homogeneous regions are combined for forming clusters with respect to common features like edge detection, intensity, the texture, threshold values, etc. From all these techniques, thresholding is popular and quite efficiently used to discriminate the objects from its backgrounds because of its simple implementation.

In bi-level thresholding approach, two homogeneous regions of total image population are found. The common techniques which are used are minimum variance, edge detection, histogram, entropy, etc. A threshold value has been found which divides images into two classes; one class contains all pixel values greater than threshold, whereas other class contains less values of threshold. Some standard bi-level thresholding approaches are noted in the literature [2, 28]. Some global entropy-based techniques are also surveyed for bi-level thresholding like Kapur, Tsallis, Wong entropies, etc. These methods are quite apparent for segmenting images into two classes, whereas multiple thresholds are required for effective extraction of objects from background for complex problems. Iterative multilevel segmentation technique based on statistics proposed by Arora et al. [3] is found in the literature as well. But those proposed approaches suffered from high time complexity problem to generate solutions as multiple threshold values. So, current research trend suggests solving multilevel complex problems by introducing optimization algorithms so that computational cost can be reduced and the outcome will be considered as optimal values in the exhaustive search space.

A meta-heuristic approach is considered as a best procedure for developing search algorithm that can find an optimal solution of a problem with limited computing capacity. Some popular optimizing techniques were surveyed in the literature, namely genetic algorithm (GA) [20], firefly (FF) algorithm [16], artificial bee colony (ABC) [13], differential evolution (DE) [25], and particle swarm optimization (PSO) [1] which produced effective optimal solutions of meta-heuristic complex problems. The limitation of the mentioned approaches found that it generated best segmented results for standard black and white images and not tested to natural color images which have been given attention in our work.

Unlike grayscale images, color image segmentation faces challenges because of its complex features. However, researchers still face challenges to segment unsupervised color images. Some notable techniques are available since last decades. A few of the common approaches are principal component analysis-based method [12], split–merge techniques [10], Markov random field models [15], mean shift clustering, hybrid method [26], texture-based color image segmentation [14], histogram-based method [8], pixel clustering [27]-based approach, etc. Further entropy-based color image thresholding methods with popular optimizers are proposed by Sarkar et al. [24]; Pare et al. [18]; Ghamsi et al. [11]; Bhandari et al. [5]; and Rajnikanth et al. [22] for effective segmentation of 3 channels of RGB images. Comparative study of nature-inspired swarm-based evolutionary computation techniques is found [9] in the literature. In the year 2017, Pare et al. have proposed multilevel color image segmentation scheme using co-occurrence matrix [19] and cross-entropy optimized by the cuckoo search algorithm [21] to produce effective results. In the very recent years, some meta-heuristic-based true color image segmentation schemes have been proposed by the various researchers [6, 17] to generate effective optimal outcomes. Bhandari et al. [4] proposed an updated DE scheme to update the constant scaling factor and crossover rate values of traditional differential evolution. Encouraged by the above-stated recent schemes, an improved DE-based thresholding approach along with fuzzy entropy (FE) has been proposed. IDE reduces the time taken and maintains enough accuracy quite effectively. Simulations have been carried out to find out the accuracy and stability of the IDE-based algorithm as compared to four recent nature-inspired optimization techniques like beta differential evolution (BDE) [4], PSO with Tsallis–Havrda–Charvat entropy [6], cuckoo search (CS) with cross-entropy [21], DE with cross-entropy [23] when applied to the 300 images of BSDS300 dataset.

Section 2 formulates the problem with multilevel FE thresholding technique. Proposed IDE algorithm has been demonstrated in Sect. 3. Simulations and comparative results of the algorithms are presented in Sect. 4. The conclusion of the paper is drawn in Sect. 5.

2 Problem Formulation

Color image multilevel thresholding generates more than two thresholds for each channel (red, green, and blue) of the image. Firstly, RGB images have been converted to CIE $L^*a^*b^*$ color space using MATLAB where L^* indicates lightness, a^* red-green, and b^* blue-yellow color, and the value of each channel lies between 0 and 255. Let I be an original color image and r_i be the histogram of the image which is the combination of three components to store the color information, i.e., $r_i = (L_i^*; a_i^*; b_i^*)$ for $i = (1, 2, \ldots L)$ where $L_i^*, a_i^*, b_i^* \rightarrow i$th intensity value. L is the number of levels (0–255) the detailed block diagram of the approach has been shown in Fig. 1.

2.1 *Fuzzy Entropy Calculated for Multiple Levels*

In classical or crisp set (say S), either of all the elements belong to S or not belong to S. But a fuzzy set states that an element can belong to partially means in between the state of 0 to 1. Let a fuzzy set F be declared as

$$F = \{(y, \mu_F(y)) \mid y \in Y\} \tag{1}$$

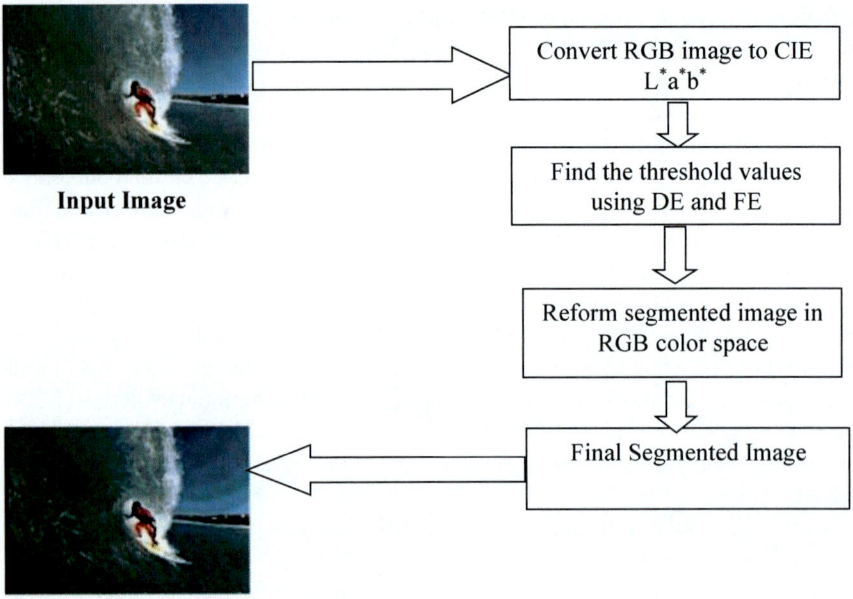

Fig. 1 Detailed block diagram of the technique

where $0 \leq \mu_F(y) \leq 1$ and $\mu_F(y)$ are called the membership function that is used to find the closeness of x to F. Trapezoidal membership function has been chosen here to calculate the membership of k segmented regions, $\mu_1, \mu_2, \ldots \mu_k$ by using $2 \times (k - 1)$ unknown fuzzy parameters. Parameters are chosen as $a_1, b_1 ..., a_k - 1, b_k - 1$, where $0 \leq a_1 \leq b_1 \leq ... \leq a_{k-1} \leq b_{k-1} \leq L - 1$. The fuzzy membership diagram can be viewed in the literature [7]. Membership function of k-level thresholding has been defined like the following:

$$\mu_1(n) = \begin{cases} 1 & n \leq a_1 \\ \frac{n-b_1}{a_1-b_1} & a_1 \leq n \leq b_1 \\ 0 & n > b_1 \end{cases} \tag{2}$$

$$\mu_{k-1}(n) = \begin{cases} 0 & n \leq a_{k-2} \\ \frac{n-a_{k-2}}{b_{k-2}-a_{k-2}} & a_{k-2} < n \leq b_{k-2} \\ 1 & b_{k-2} < n \leq a_{k-2} \\ \frac{n-b_{k-1}}{a_{k-1}-b_{k-1}} & a_{k-1} < n \leq b_{k-1} \\ 0 & n > b_{k-1} \end{cases} \tag{3}$$

$$\mu_k(n) = \begin{cases} 1 & n \leq a_{k-1} \\ \frac{n-a_k}{b_k-a_k} & a_{k-1} < n \leq b_{k-1} \\ 1 & n > b_{k-1} \end{cases} \tag{4}$$

The maximum fuzzy entropy for each of k-level segmentation can be expressed as:

$$\left.\begin{aligned} H_1 &= -\sum_{i=0}^{L-1} \tfrac{r_i*\mu_1(i)}{R_1} * ln(\tfrac{r_i*\mu_1(i)}{R_1}), \\ H_2 &= -\sum_{i=0}^{L-1} \tfrac{r_i*\mu_2(i)}{R_2} * ln(\tfrac{r_i*\mu_2(i)}{R_2}), \\ &\quad . \\ H_k &= -\sum_{i=0}^{L-1} \tfrac{r_i*\mu_k(i)}{R_k} * ln(\tfrac{r_i*\mu_k(i)}{R_k}), \end{aligned}\right\} \tag{5}$$

where $R_1 = \sum_{i=0}^{L-1} r_i * \mu_1(i),\ R_2 = \sum_{i=0}^{L-1} r_i * \mu_2(i), \ldots,\ R_k = \sum_{i=0}^{L-1} r_i * \mu_k(i)$

The optimal value of the function can be achieved by maximizing the fuzzy entropy as follows:

$$\psi(a_1, b_1, ...a_{k-1}, b_{k-1}) = \max([H_1(t) + H_2(t)]f + \cdots + H_k(t)]) \tag{6}$$

To make computational technique easier, two thresholds have been introduced as $t_0 = 0$ and $t_k = L - 1$ where $t_0 < t_1 < .... < t_{k-1} < t_k$. The following formula helps to generate $(k - 1)$ number of thresholds.

$$t_1 = \frac{(a1 + b1)}{2}, t_2 = \frac{(a2 + b2)}{2}, \ldots, t_{k-1} = \frac{(a_{k-1} + b_{k-1})}{2} \tag{7}$$

The time complexity of multilevel thresholding is as high as $O(L^{k-1})$ and will increase to the increment of segmentation level (k). So, here DE can be used to optimize Eq. 6 so that time complexity gets reduced and algorithm generates best near-optimal thresholds.

3 Proposed Optimization Model

Proposed optimization model is based on the proposal of an improved differential evolution (IDE) algorithm. First and foremost, the standard DE algorithm has been defined.

3.1 Differential Evolution (DE)

Differential evolution is simple and widely accepted heuristic algorithm of current research interests, proposed by Storn and Price in 1997. The algorithm can be divided into four parts like initialization, mutation, crossover, and selection. The algorithm stops when any termination criterion is met like exhaustion of maximum functional evaluations, and then the loop of finding of new generation technique stops automatically. Functional evaluations or fitness values help to get solutions of a particular problem.

The jth individual parameter vector of the population at generation G is an M-dimensional vector. It contains a set of M optimization parameters (decision variables):

$$\overrightarrow{X_{j,G}} = [X_{j,1,G}, X_{j,2}, X_{j,3,G} \dots X_{j,M,G}] \tag{8}$$

The search space with maximum–minimum bounds can be declared as: $\overrightarrow{X_{\max}} = [x_{\max,1}, x_{\max,2}, \dots x_{\max,M}]$ and $\overrightarrow{X_{\min}} = [x_{\min,1}, x_{\min,2}, \dots x_{\min,M}]$. Now the ith component of the jth individual can be declared by the following:

$$\overrightarrow{x_{j,i}}(0) = x_{\min,i} + rand_{j,i}(0, 1).(x_{\max,i} - x_{\min,i}), \tag{9}$$

where $i \in \{1, 2, ...M\}$ and $rand_{j,i}(0, 1)$ are uniformly distributed random number in (0,1). Population numbers $\overrightarrow{X_{j,G}}$ of each generation can be changed with the introduction of a mutant or donor vector $\overrightarrow{Y_{j,G}}$. This donor or mutant vector is used to distinguish the different DE schemes. In DE/rand/1 scheme (one of the modified DE versions), three other parameter vectors (say $p1$, $p2$, and $p3$th vectors such that $p1$, $p2$, $p3$ $\in$[1,P], where P denotes the population size and $p1 \neq p2 \neq p3$) are chosen randomly to create a donor vector $\overrightarrow{Y_{j,G}}$ for each jth member. Now donor vector $\overrightarrow{Y_{j,G}}$ is obtained among one of the three vectors where scaling differences between two vectors are also considered. So, the mutation process for an ith component of the jth vector can be defined like below:

$$\overrightarrow{Y_{j,i,G}} = X_{p1,i,G} + F.(X_{p2,i,G} - X_{p3,i,G}), \tag{10}$$

where F is a scalar number (usually called scale factor of DE and often lies between [0.4,2]) and is used to control the amplification between different vectors. According to Storn and Price [?], DE follows two types of crossover schemes named exponential and binomial crossover. A binomial scheme, chosen in our approach, uses a control parameter C (known as crossover rate for DE) to perform the binomial crossover for each of the M variables. So for each corresponding vector of $\overrightarrow{X_j}$, a trial vector $\overrightarrow{S_{j,G}}$ is created like the following:

$$s_{j,i,G} = \begin{cases} y_{j,i,G}, & \text{if } rand_{j,i}(0,1) \leq C \| i = pn(j) \\ X_{j,i,G}, & \text{otherwise} \end{cases} \tag{11}$$

where $i = 1, 2 \ldots M$ and $rand_{j,i}(0, 1) \in [0, 1]$ is the random number generator for the ith evaluation and pn(j)∈[1,2, …M] is randomly chosen integer which specifies that $\overrightarrow{S_{j,G}}$ gets at least one component from $\overrightarrow{X_{j,G}}$. Next selection process is evaluated to determine which vector among target and trial will survive in the next generation. Now the comparison is carried out between target and trial vectors to select which one produces better values at next generation. If trial overcomes the challenge of its parent vector, then it will sustain in next generation; otherwise, parent retains its position in the population as defined below:

$$\overrightarrow{X_{j,G+1}} = \begin{cases} \overrightarrow{S_{j,G}}, & \text{if } f(\overrightarrow{S_{j,G}}) \leq f(\overrightarrow{X_{j,G}}) \\ \overrightarrow{X_{j,G}}(t), & \text{if } f(\overrightarrow{S_{j,G}}) > f(\overrightarrow{X_{j,G}}) \end{cases} \tag{12}$$

where function $f()$ has to be maximized.

3.2 Proposed Improved DE

In the conventional DE, the convergence speed may lack due to the constant scaling factor (F) and crossover rate (Cr) of DE. The standard value of F generally lies between 0.4 and 2. Bhandari et al. [4] proposed a beta differential evolution scheme where probability distribution has been applied using betarand() function to update the scaling factor (F) and crossover rate (Cr) dynamically for improving the speed of convergence. Encouraged by that scheme, we are proposing an idea where value of scale factor will be generated randomly using the below equation:

$$F = randg(2 * rand) * (rand - rand) \tag{13}$$

where $randg()$ returns a scalar random value chosen from a gamma distribution with unit scale and shape and $rand$ returns a single uniformly distributed random number in the interval (0,1). So, $(rand - rand)$ will not result to 0; rather, it will provide different subtraction values for each iteration. This approach helps to find population diversity among stochastic search mechanism. Along with the random generation of scaling values, we are also proposing to decrease the crossover rate linearly with time variations. The range of Cr will lie between 0.5 and 1 and will be indicated by $Cr_{\max} = 1.0$ and $Cr_{\min} = 0.5$. The linear decreasing of Cr rate will specify that the parent vector's components will be updated by the child vector as per Eq. 11. But in the closing stages of the optimization approach where Cr value will reach close to 0.5, more components of the parent vector will be inherited to generate the optimal value search space. As a result of that, the search space for finding global optima will be reduced and hence the algorithm will result in speedy convergence. The variation of the Cr values with respect to times can be expressed by the below formula:

$$Cr = (Cr_{\max} - Cr_{\min}) * (\text{MaxIter} - \text{itr})/\text{MaxIter} \tag{14}$$

where itr specifies the number of present iterations and $MaxIter$ mentions the maximum allowable iterations. The time complexity of standard DE algorithm for G no. of iterations is $O(P * M * G)$ in exhaustive search space [24]. In the fuzzy entropy-based multilevel thresholding problem, $M = 2 * (k - 1)$ so the revised computational cost will be $O(P * (k - 1) * G)$ which is a linear parameter and obviously less than the exponential cost which we calculated earlier.

4 Experimental Results

4.1 Experimental Setup

The experiment is carried out with combination of fuzzy entropy and DE optimization algorithm. Fuzzy entropy is measured in Eq. 6; later, DE optimization algorithm is applied on it. The threshold values of each channel have been calculated by maximizing the objective function and finally reformed it to get the resultant RGB segmented images for k-level problems. The detailed flow diagram can be seen in Fig. 2.

Here, MATLAB R2018a workstation has been used to perform the simulation in personal computer. The core i5 3.2 GHz processor along with RAM of 4 GB is used. The parametric setup of BDE [4], PSO [1], CS [5], and DE [25] has been chosen by following the guidelines given in the above literature and noted in Table 3. Best mean (f_m) among 100 independent runs is calculated where each run was performed till the completion of M $\times$ 1000 number of iterations. Now, $M = 2*(k - 1)$ where k is the segmentation level and 2 fuzzy parameters. High segmentation levels like 7 and 10 have been chosen in the experiment so that better outcomes can be provided to the reader. Evaluation of statistical measurements has been carried out in the form of f_m, standard deviation f_{std}, and computing time f_t, whereas the visual quality of segmented color images has been measured using the popular measurements, namely FSIM [29], PSNR, and SSIM [16].

4.2 Performance Analysis

Berkley Segmentation Dataset (BSDS300) is chosen in our experiment. The dimension of the images in the set is 484$\times$321. The data can be freely downloaded from the Web site link **https://www2.eecs.berkeley.edu/Research/Projects/CS/vision/bsds/**. Four popular optimization algorithms named BDE [4], PSO [6], CS [21], and DE [23] have been chosen for comparing the results. Table 1 shows the computational time in seconds, best mean function values, and f_{std} of six selected images from the dataset

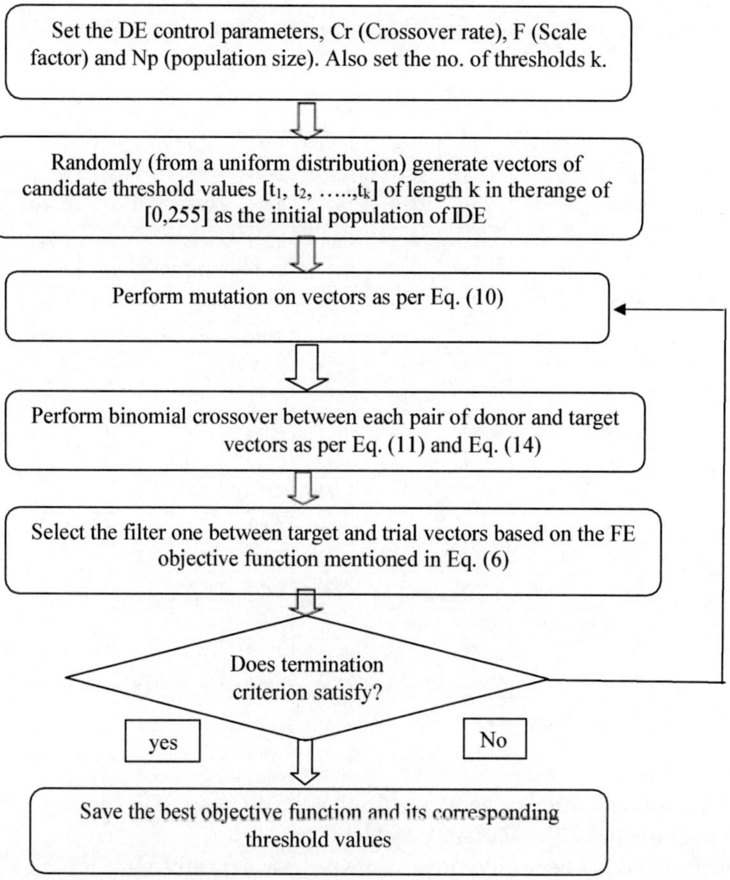

Fig. 2 Flowchart of the proposed approach

for all the optimization algorithms using fuzzy entropy. The best values produced by our approach for the levels mentioned are boldly marked in the table. High mean objective values in quick time ensure the better working of our algorithm for all the levels. The dynamic update of the F and Cr of IDE has taken less time in comparison with conventional DE also shown in the above table. The IDE-fuzzy algorithm is found as more stable because standard deviation values tend to 0 for all the levels in our algorithm. BDE is more stable than PSO, CS, and DE but suffers more in terms of premature convergence. Comparative convergence and computation time graphs of “38,092.jpg” from BSDS300 of two different levels 7 and 10 have been shown in Fig. 3. In 7th and 10th Lv, the convergence rate of IDE is at 10,950 and 13,700 FEs, whereas BDE, PSO, CS, and DE converge at 11,400, 11,600, 12,000, and 13,100 for 7-level and 14,300, 14,700, 15,000, and 15,600 for 10-level, respectively. Time plots have been drawn to show the fastness of the algorithm. IDE takes 35 s to compute the results from 7th to 10th Lv, whereas remaining algorithms consume much time

Table 1 Computing time (t), best mean objective value (f_m), and standard deviation (f_{std}) comparison between IDE, BDE, PSO, CS, and DE

Im		7-level					10-level				
		IDE	BDE	PSO	CS	DE	IDE	BDE	PSO	CS	DE
1	t	**3.619**	3.890	4.782	4.895	5.053	**5.191**	5.407	7.298	7.347	7.482
	f_m	**20.718**	20.698	20.568	20.358	20.204	**27.457**	27.101	26.904	26.621	26.257
	f_{std}	**0.000**	0.000	0.678	0.689	0.583	**0.000**	0.000	0.559	0.360	0.488
2	t	**3.628**	4.098	4.712	4.881	5.101	**5.271**	6.667	7.897	7.942	8.082
	f_m	**20.729**	20.606	20.575	20.365	20.213	**27.464**	27.202	26.924	26.632	26.276
	f_{std}	**0.000**	0.000	0.677	0.692	0.595	**0.000**	0.000	0.565	0.370	0.498
3	t	**3.679**	4.202	4.721	4.890	5.045	**5.012**	6.666	7.457	7.589	7.731
	f_m	**20.779**	20.646	20.589	20.382	20.294	**27.487**	27.206	26.994	26.741	26.387
	f_{std}	**0.000**	0.000	0.679	0.679	0.585	**0.000**	0.000	0.547	0.368	0.482
4	t	**3.645**	4.102	4.757	4.950	5.184	**5.181**	6.989	7.321	7.478	7.588
	f_m	**20.828**	20.707	20.678	20.552	20.409	**27.627**	27.106	26.994	26.731	26.455
	f_{std}	**0.000**	0.000	0.575	0.589	0.533	**0.000**	0.000	0.519	0.350	0.428
5	t	**3.542**	4.192	4.737	4.967	5.109	**5.221**	6.770	7.026	7.282	7.367
	f_m	**20.618**	20.578	20.515	20.408	20.324	**27.447**	27.109	26.914	26.611	26.327
	f_{std}	**0.000**	0.000	0.668	0.679	0.593	**0.000**	0.000	0.549	0.380	0.478
6	t	**3.542**	4.098	4.837	4.967	5.109	**5.331**	6.198	7.151	7.356	7.478
	f_m	**20.738**	20.667	20.588	20.387	20.314	**27.477**	27.222	27.004	26.821	26.557
	f_{std}	**0.000**	0.000	0.608	0.629	0.593	**0.000**	0.000	0.539	0.390	0.468

than this. It is found from the above results that the proposed IDE converges fast and does not get trapped in local optima problem.

Now the focus has been given to measure the visual quality of the results. Figure 4 displays the original and segmented images obtained by using fuzzy entropy with various optimization algorithms to show the visual clarity achieved by IDE. Now SSIM, FSIM, and PSNR have been calculated as below: SSIM:

$$\text{SSIM}(x, y) = \frac{(2\mu_I \mu_{I'} + C_1)(2\sigma_{II'} + C_2)}{(\mu_I^2 \quad + \mu_{I'}^2 + C_1)(\sigma_I^2 + \sigma_{I'}^2 + C_2)}$$

$$\text{for color image, SSIM} = \sum_c \text{SSIM}(x^c, y^c) \tag{15}$$

where μ_I and $\mu_{I'}$ represent mean of both the images, σ_I and $\sigma_{I'}$ represent standard deviation , $\sigma_{II'}$ is cross-correlation, $C1$, $C2$ are constants, and c represents color.

FSIM is given in [29]:

$$\text{FSIM}_c = \frac{\sum_{x\in\Omega} S_L(x).[S_C(x)]^\lambda.PC_m(x)}{\sum_{x\in\Omega} PC_m(x)} \tag{16}$$

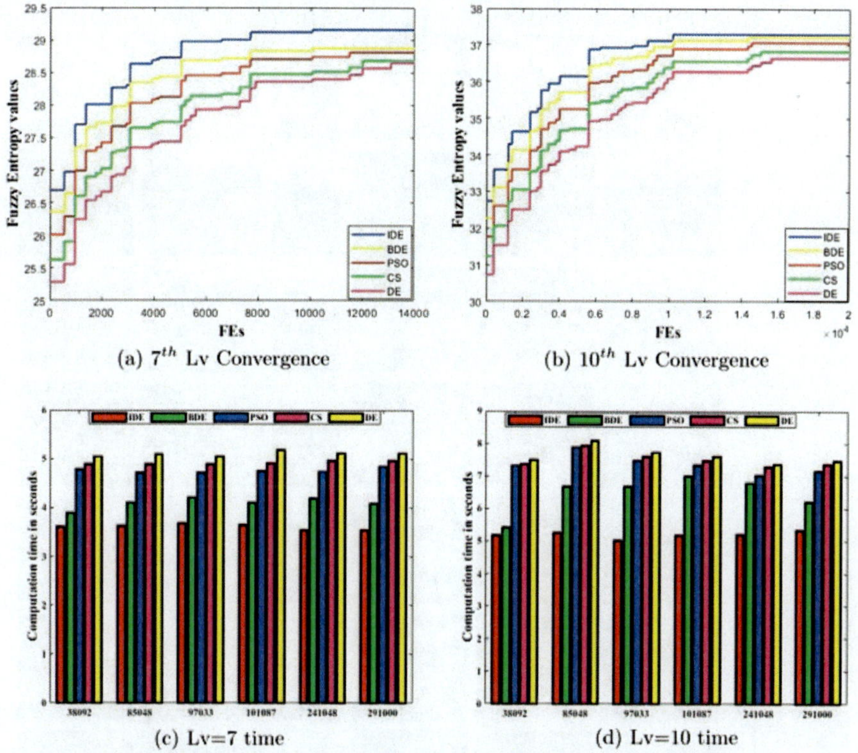

(a) 7th Lv Convergence (b) 10th Lv Convergence

(c) Lv=7 time (d) Lv=10 time

Fig. 3 Convergence and time plots

PSNR:

$$\text{PSNR} = 10 \times \log_{10}\left(\frac{255^2}{\text{RMSE}}\right)(dB)$$

$$\text{where RMSE } = \frac{1}{R \times C}\sum_{i}^{R}\sum_{j}^{C}\{I(i,j) - I'(i,j)\}^2 \quad (17)$$

where I is original image, $I^{'}$ is segmented image, $R \times C$ is total size, and RMSE is root mean square error. In Table 2, corresponding PSNR, SSIM, and FSIM are noted. The proposed optimization approach along with fuzzy entropy generates best results as compared to the other techniques. It is noted that $-1 \leq FSIM, SSIM \leq +1$. -1 tells worst result of segmentation, and +1 value cannot discriminate two images. It can be seen that the values of the metrics are closer to 1 for high level for the proposed approach and values are increasing proportionally to the level. The outcomes of our approach are boldly marked in the table. Similarly, it is noted that high PSNR values of the proposed algorithm outperform other state-of-the-art approaches (Table 2).

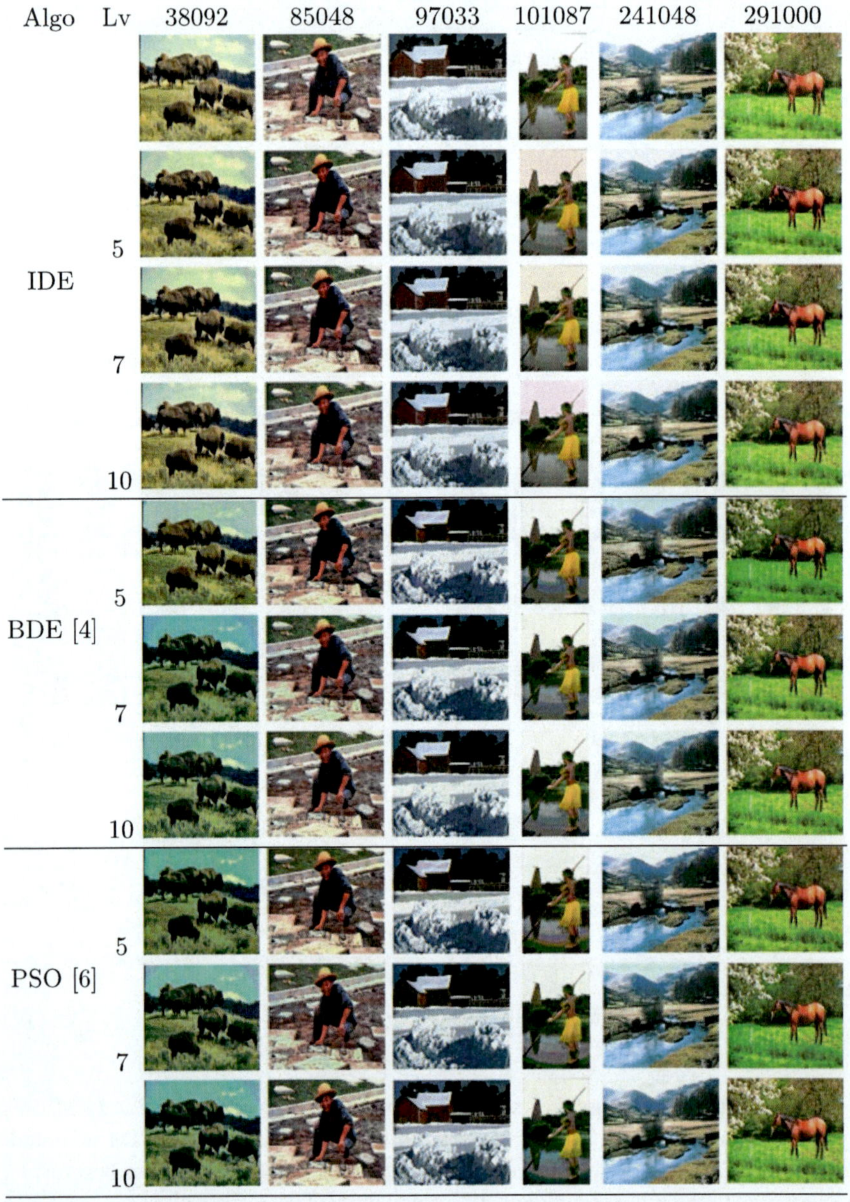

Fig. 4 Segmented images with different optimizers using fuzzy entropy

Fig. 4 (continued)

5 Conclusion and Future Work

Histogram-based color image segmentation technique has been successfully implemented. Fuzzy entropy-based approach combined with IDE delivers outstanding results in the reasonable amount of computational time. IDE outperforms four widely known derivative-free nature-inspired meta-heuristic global optimizers BDE, PSO, CS, and DE. Image quality assessment metrics SSIM, FSIM, and PSNR doubtlessly established IDE as an effective and robust optimizer. However, more improved version of DE with upgraded variants can be proposed to generate better performance. In the future, region merging-based technique can be applied with updated DE parameters to obtain more distinguishable segmented regions of color images. Finally, some other image segmentation metrics can be tested to prove the efficacy of the algorithm.

Table 2 Comparative results of PSNR, SSIM, and FSIM values between IDE, BDE, PSO, CS, and DE aided with fuzzy entropy

Im		7-level					10-level				
		IDE	BDE	PSO	CS	DE	IDE	BDE	PSO	CS	DE
1	PSNR	**18.657**	17.703	17.380	17.267	15.085	**19.919**	18.049	17.898	17.647	17.482
	SSIM	**0.777**	0.666	0.691	0.675	0.520	**0.830**	0.805	0.799	0.711	0.627
	FSIM	**0.852**	0.834	0.821	0.774	0.789	**0.869**	0.857	0.834	0.802	0.798
2	PSNR	**18.757**	18.202	17.680	17.467	15.585	**19.988**	18.777	18.098	17.777	17.687
	SSIM	**0.785**	0.706	0.697	0.684	0.570	**0.839**	0.828	0.809	0.761	0.727
	FSIM	**0.844**	0.837	0.831	0.770	0.759	**0.872**	0.865	0.842	0.792	0.778
3	PSNR	**18.877**	18.208	17.790	17.677	15.885	**20.098**	19.096	18.374	17.796	17.889
	SSIM	**0.788**	0.743	0.707	0.699	0.592	**0.848**	0.830	0.811	0.775	0.737
	FSIM	**0.854**	0.848	0.837	0.774	0.769	**0.880**	0.865	0.847	0.799	0.785
4	PSNR	**18.746**	17.989	17.682	17.547	15.573	**19.972**	18.789	18.058	17.765	17.661
	SSIM	**0.765**	0.732	0.675	0.659	0.566	**0.837**	0.818	0.801	0.760	0.725
	FSIM	**0.832**	0.824	0.810	0.764	0.752	**0.870**	0.858	0.841	0.791	0.768
5	PSNR	**18.766**	17.900	17.685	17.469	15.579	**20.085**	19.376	18.097	17.769	17.682
	SSIM	**0.781**	0.702	0.690	0.682	0.575	**0.832**	0.822	0.800	0.760	0.721
	FSIM	**0.840**	0.828	0.801	0.779	0.767	**0.870**	0.856	0.844	0.794	0.779
6	PSNR	**18.762**	17.978	17.682	17.571	15.594	**19.999**	19.200	18.234	17.887	17.786
	SSIM	**0.789**	0.705	0.699	0.679	0.575	**0.833**	0.812	0.804	0.769	0.738
	FSIM	**0.840**	0.832	0.821	0.788	0.753	**0.860**	0.851	0.844	0.791	0.779

Table 3 Parameters set up for the chosen algorithms IDE, DE, PSO, BDE, and CS

Name of the algorithm	Parameters	Values
IDE	NP	$10\times M$
	F	As per Eq. 13
	Cr	As per Eq. 14
DE	NP	10
	F	0.5
	Cr	0.5
PSO	w_{max}, w_{min}	0.4, 0.1
	C_1, C_2	2
	Size of swarm	100
PSO	NP	$10\times$ M
	F	0.5
	Cr	0-0.2
	No. of objectives	1
	No. of constraints	0
	No. of decision variables	4
CS	Nests	25
	Iterations	3000
	Step size (α)	1
	Mutation probability (p_α)	0.25
	Scaling factor (β)	1.5

References

1. Akay B (2013) A study on particle swarm optimization and artificial bee colony algorithms for multilevel thresholding. Appl Soft Comput 13(6):3066–3091
2. Arifin AZ, Asano A (2006) Image segmentation by histogram thresholding using hierarchical cluster analysis. Pattern Recogn Lett 27(13):1515–1521
3. Arora S, Acharya J, Verma A, Panigrahi PK (2008) Multilevel thresholding for image segmentation through a fast statistical recursive algorithm. Pattern Recogn Lett 29(2):119–125
4. Bhandari AK (2018) A novel beta differential evolution algorithm-based fast multilevel thresholding for color image segmentation. Neural computing and applications, pp 1–31
5. Bhandari AK, Kumar A, Chaudhary S, Singh GK (2016) A novel color image multilevel thresholding based segmentation using nature inspired optimization algorithms. Expert Syst Appl 63:112–133
6. Borjigin S, Sahoo PK (2019) Color image segmentation based on multi-level Tsallis-Havrda-charvát entropy and 2d histogram using PSO algorithms. Pattern Recogn 92:107–118
7. Chakraborty R, Sushil R, Garg M (2019) Hyper-spectral image segmentation using an improved pso aided with multilevel fuzzy entropy. Multimedia Tools Appl, pp 1–37
8. Chen S, Cao L, Wang Y, Liu J, Tang X (2010) Image segmentation by map-ml estimations. IEEE Trans Image Process 19(9):2254–2264
9. Chouhan SS, Kaul A, Singh UP (2018) Soft computing approaches for image segmentation: a survey. Multimedia Tools Appl 77(21):28483–28537
10. Garcia-Ugarriza L, Saber E, Amuso V, Shaw M, Bhaskar R (2008) Automatic color image segmentation by dynamic region growth and multimodal merging of color and texture information. In: IEEE international conference on acoustics, speech and signal processing. ICASSP 2008. IEEE, pp 961–964
11. Ghamisi P, Couceiro MS, Martins FM, Benediktsson JA (2014) Multilevel image segmentation based on fractional-order Darwinian particle swarm optimization. IEEE Trans Geosci Remote Sens 52(5):2382–2394
12. Han Y, Feng XC, Baciu G (2013) Variational and pca based natural image segmentation. Pattern Recogn 46(7):1971–1984
13. Karaboga D, Gorkemli B, Ozturk C, Karaboga N (2014) A comprehensive survey: artificial bee colony (abc) algorithm and applications. Artif Intell Rev 42(1):21–57
14. Krinidis M, Pitas I (2009) Color texture segmentation based on the modal energy of deformable surfaces. IEEE Trans Image Process 18(7):1613–1622
15. Mignotte M (2008) Segmentation by fusion of histogram-based *k*-means clusters in different color spaces. IEEE Trans Image Process 17(5):780–787
16. Naidu M, Kumar PR, Chiranjeevi K (2017) Shannon and fuzzy entropy based evolutionary image thresholding for image segmentation. Alexandria Eng J
17. de Oliveira PV, Yamanaka K (2018) Image segmentation using multilevel thresholding and genetic algorithm: An approach. In: 2018 2nd international conference on data science and business analytics (ICDSBA). IEEE, pp 380–385
18. Pare S, Bhandari A, Kumar A, Singh G (2017) A new technique for multilevel color image thresholding based on modified fuzzy entropy and lévy flight firefly algorithm. Comput Electr Eng
19. Pare S, Bhandari AK, Kumar A, Singh GK (2017) An optimal color image multilevel thresholding technique using grey-level co-occurrence matrix. Expert Syst Appl 87:335–362
20. Pare S, Bhandari AK, Kumar A, Singh GK, Khare S (2015) Satellite image segmentation based on different objective functions using genetic algorithm: a comparative study. In: 2015 IEEE international conference on digital signal processing (DSP). IEEE, pp 730–734
21. Pare S, Kumar A, Bajaj V, Singh GK (2017) An efficient method for multilevel color image thresholding using cuckoo search algorithm based on minimum cross entropy. Appl Soft Comput 61:570–592
22. Rajinikanth V, Couceiro M (2015) Rgb histogram based color image segmentation using firefly algorithm. Procedia Comput Sci 46:1449–1457

23. Sarkar S, Das S, Chaudhuri SS (2015) A multilevel color image thresholding scheme based on minimum cross entropy and differential evolution. Pattern Recogn Lett 54:27–35
24. Sarkar S, Das S, Chaudhuri SS (2016) Hyper-spectral image segmentation using rényi entropy based multi-level thresholding aided with differential evolution. Expert Syst Appl 50:120–129
25. Sarkar S, Paul S, Burman R, Das S, Chaudhuri SS (2014) A fuzzy entropy based multi-level image thresholding using differential evolution. In: International conference on swarm, evolutionary, and memetic computing. Springer, pp 386–395
26. Tan KS, Isa NAM (2011) Color image segmentation using histogram thresholding-fuzzy c-means hybrid approach. Pattern Recogn 44(1):1–15
27. Yu Z, Au OC, Zou R, Yu W, Tian J (2010) An adaptive unsupervised approach toward pixel clustering and color image segmentation. Pattern Recogn 43(5):1889–1906
28. Zaitoun NM, Aqel MJ (2015) Survey on image segmentation techniques. Procedia Comput Sci 65:797–806
29. Zhang L, Zhang L, Mou X, Zhang D (2011) Fsim: a feature similarity index for image quality assessment. IEEE Trans Image Process 20(8):2378–2386

Chapter 4
Clustered Fault Repairing Architecture for 3D ICs Using Redundant TSV

Sudeep Ghosh, Mandira Banik, Moumita Das, Tridib Chakraborty, Chowdhury Md. Mizan, and Arkajyoti Chakraborty

Abstract Through silicon via (TSVs) base 3D integrated circuit (3D IC) has become most emerging technology in semiconductor industry. TSVs-based 3D-IC has some advantages like low foot-print area, less power requirement, small interconnection length, etc. The major concern is manufacturing defect in for 3-DIC. There may have some manufacturing defect in TSVs. A single fault can destroy the total chip. So, it is required to repair faulty TSV to make a chip functional. An effective repairing method is required. The one solution would be use of redundant TSVs the reroute the signal. As of now, there are very few works exists to address problem of clustered fault for irregularly distributed TSVs all over the chip. Some works exist on regularly distributed TSVs but very few in irregularly distributed TSVs. In our work, we have proposed a method to form groups among functional and redundant TSVs and make connection between functional and redundant TSVs in an innovative way that we can repair clustered fault. Also, we tried to use minimum number multiplexer (MUXs) so that area overhead will be reduced.

Keywords TSVs · 3D IC · MUXs · Dependency

S. Ghosh (✉) · M. Banik · M. Das · T. Chakraborty · C. Md. Mizan · A. Chakraborty
Guru Nanak Institute of Technology, Kolkata, India
e-mail: sudeep.ghosh@gnit.ac.in

M. Banik
e-mail: mandira.banik@gnit.ac.in

M. Das
e-mail: moumita.das@gnit.ac.in

T. Chakraborty
e-mail: tridib.chakraborty@gnit.ac.in

C. Md. Mizan
e-mail: chowdhurymd.mizan@gnit.ac.in

J. K. Mandal et al. (eds.), *Proceedings of International Conference on Innovations in Software Architecture and Computational Systems*, Studies in Autonomic, Data-driven and Industrial Computing, https://doi.org/10.1007/978-981-16-4301-9_4

1 Introduction

In current, the territory of electronic gadgets is contracting step by step. Therefore, the development of 3D-IC is proved to be most significant field in semiconductor industry. The 3D-IC is made by various dies stacked individually to frame distinctive layer and all the layer are vertically connected by TSVs. So TSVs are acted as vertical connector of stacked dies. There are few advantages of 3D IC compared to 2D IC like utilization of less power, small impression, diminished interconnection length, rapid, etc. [1, 2]. The main concern on TSV-based 3D-IC is dependability. Since there may be different kinds of TSVs absconds because of complicated assembling procedure of 3D-IC [3]. Any solitary defective TSV present in a chip may crush the entire chip. Thus, recuperation technique is needed to make a sustainable chip.

An excess TSV may be utilized to replace a flawed TSV to avert the signal from one layer to another layer. In this way, a chip will be in functional condition in spite of having flawed TSVs in that chip. Gathering of blemished and excess TSVs will be an alluring answer for actualize functional TSVs engineering. In [4–6] creators examined various strategies for gathering of useful and excess TSVs. In [7], bunches have been shaped relying upon accessible number of multiplexers. In the work [8, 9], creators talked about the engineering for consistently conveyed TSVs in 3D-IC. Be that as it may, by and large TSVs are circulated sporadically everywhere on the chip. In this work, we conferred a strategy to replace a defective TSV by excess TSV with the goal that the chip will be in working state. To keep the small chip region, we tried to used small number of MUXs utilized for recuperation. Thus, we center to give higher fix ability and furthermore attempt to utilize least number of MUXs with the goal that complete territory for MUXs will be least. We have utilized sign moving technique between utilitarian TSVs and direct association among practical and functional TSVs.

The remainder of the paper is sorted out as pursues. Segment 2 portrays the past works which shows the methods to recuperation flawed TSVs. In Segment 3, we depict our motivation and formulated the problem. Segment 4 expounds discussed strategy and a model is exhibited in Segment 5. Segment 6 conferred exploratory outcomes and comparison with previous results, lastly Segment 7 finishes up this paper.

2 Prior Works

In semiconductor industry flaw compassionate 3D-IC dependent is a rising zone. One blemished TSV can crush an entire chip, so a defective TSV plays a significant role to make the chip usefulness. In the work [3], creators had talked about different TSVs issues. Different TSVs testing strategies which distinguish flawed TSVs are likewise were proposed in the work [10, 11], and the creators exhibited probing techniques of reducing TSVs testing time.

The technique of a TSV. repetition 3D memory is displayed in the work [12]. In the work [13], a gathering thought is explained where two excess TSVs and four functional TSVs are formed a group. So, it is possible to repair only two blemished TSVs.

In the work [6, 7], the idea of different reliance is explained. Creators suggested a strategy to make gatherings of utilitarian and functional TSVs so that at least two gatherings are covered and covered useful TSVs could be bolstered by excess TSVs exist in various gatherings.

In this work, we have proposed a non-consistently appropriated TSV engineering and a strategy to create interconnection between active TSVs and excess TSVs with the goal of unwavering quality of the IC is improved.

3 Motivation and Problem Formulation

In semiconductor industry flaw compassionate 3D-IC dependent is a rising zone. One blemished TSV can crush an entire chip, so a defective TSV plays a significant role to make the chip usefulness. In the work [3], creators had talked about different TSVs issues. Different TSVs testing strategies which distinguish flawed TSVs are likewise were proposed in the work [10, 11], and the creators exhibited probing techniques of reducing TSVs testing time.

The possibility of various conditions is assuming a crucial job to design a chips utilitarian. Reliance of the gathering determined the quantity of useful TSVs upheld by a functional TSV and aggregate of the conditions of all gatherings is defined as dependency. Covering of gatherings builds the complete reliance; consequently, recovery of flawed TSVs is improved. In the work [7], all gatherings are comparative so bunch proportion is same for all gatherings and large number of multiplexers would be required to execute this engineering. Likewise, in the event that a gathering contains more than one functional TSV, at that point absolute wire length will be expanded.

On the off chance that we limit the quantity of multiplexers, at that point region overhead of recuperation system will be reduced. Inspired by all these things the we propose a recuperation technique to fix flawed TSVs by utilizing minimum number of MUXs.

The objective of this paper is to make legitimate gatherings of practical and functional TSVs so that it utilizes least number of MUXs and thus expands the general reliance. In this way, the issue proclamation could be characterized as pursues:

> Suppose We are given the number of functional TSVs denoted by (N) and number of excess TSVs denoted by (R) with their known location, we have to find the technique to form groups in such a way that utilization of MUXs will be less ads and also it can repair clustered fault.

We presented a method to solve above problem.

4 Proposed Method

We proposed a heuristic algorithm to solve the problem as we described in previous section. Our method contains two parts. These are (1) technique to make connection between all active TSVs, and (2) Technique to form group active and excess TSVs.

4.1 Technique to of Making Interconnection Among All Active TSVs

As we are not aware of the actual size of a chip and the placement of TSVs. So, we arbitrarily produce directions of active and excess TSVs in standardized region (running from 0 to 1) with the goal that it may be mapped to any realized chip size.

The steps which are required to make connection among active and excess TSVs:

1. The primary useful TSV will be the nearest of (0, 0) co-ordinate.
2. Make connection this TSV with its closest utilitarian TSV. relying upon separation.
3. Now we are having one edge with two vertices expecting the TSV positions as hubs. Discover the closest utilitarian TSVs from both vertices. Pick that TSV of most brief separation and interface them. So, we are having length of 2.
4. Now we are having 3 useful TSV which are associated. Presently select the closest utilitarian TVSs of the both parts of the bargains and associate with the closest one.
5. Repeat the step 4 for remaining TSVs until the whole TSVs are associated. Subsequently, all the utilitarian TSVs are associated in diving request contingent upon their good ways from (0, 0) arrange and we get a way of length (N-1) for "N" quantities of practical TSVs.

4.2 Techniques to Make Connection Between Functional TSVs and Redundant TSVs

We placed a redundant TSV in each group. And each excess TSVs will be directly connected to one functional TSVs in each group. If there are "m" number groups so there should be similar number of excess TVS, i.e., "m" number of excess TSV. And each group there will be one redundant TSV and (m-1) number of functional TSVs.

4.3 Required MUXs Number Calculation

To create interconnection among all utilitarian TSVs, we required to interface every useful TSV to its previous and next useful TSVs. So, it is required one 3-to-1 MUX to make interconnection for each useful TSV. Be that as it may, first TSV is associated with its next useful TSV and the last practical TSV is associated with its previous TSV. So, we need one 2-to-1 MUXs for every starting and ending practical TSV.

Thus, absolute number of MUXs required is:

$[N-2]$ 3-to-1 MUXs and two 2-to-1 MUXs required to create connection between all practical TSVs. We can use one 3-to-1 MUXs by utilizing two 2-to-1 MUXs, so absolute number of expected 2-to-1 MUXs is:

$$[2(N-2)+2] \tag{1}$$

MUXs count for creating association among practical and excess TSV:

Let R number of active TSVs are associated with excess TSVs (utilizing technique B). To execute this, we required one 2-to-1 MUXs for every useful TSVs. So, total number of required 2-to-1 MUXs is R. Consequently, total 2-to-1 number of MUXs required is:

$$[2(N-2)+2+R] = [2(N-1)+R] \tag{2}$$

4.4 Repairing Path

A blemished TSVs can be replaced by an active TSVs by following techniques:

1. First, we need to check whether it is legitimately associated with excess TSV or not. In the event that it is straightforwardly associated with excess TSV, at that point it will redundant by functional TSV.
2. If it is not associated with the excess TSV then its past and previous TSVs must be associated with the functional TSV as each redundant TSVs are directly associated with functional TSVs. At that point, it is required to check initially its previous TSV whether it is now utilized by another flawed TSV or not. In the event that is not being used, at that point signal will be moved through its previous TSV generally signal will be moved through its next useful TSV.

5 Illustrative Example

Figures 1, 2, and 3 show the connection between functional TSVs redundant TSVs. Each of the existing redundant architectures has differing characteristics. The switching and shifting methods are used to repair a few faults. However, they are not suitable for clustered faults because of their structural limitations. Therefore, ring

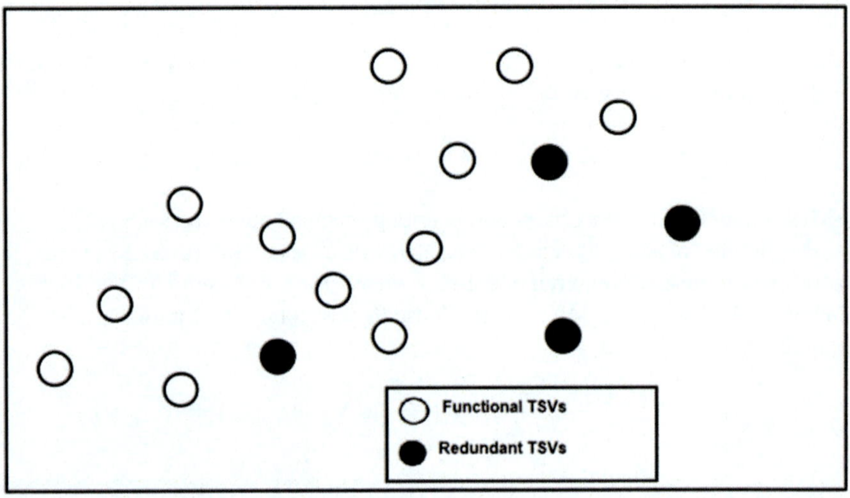

Fig. 1 Functional & redundant TSVs

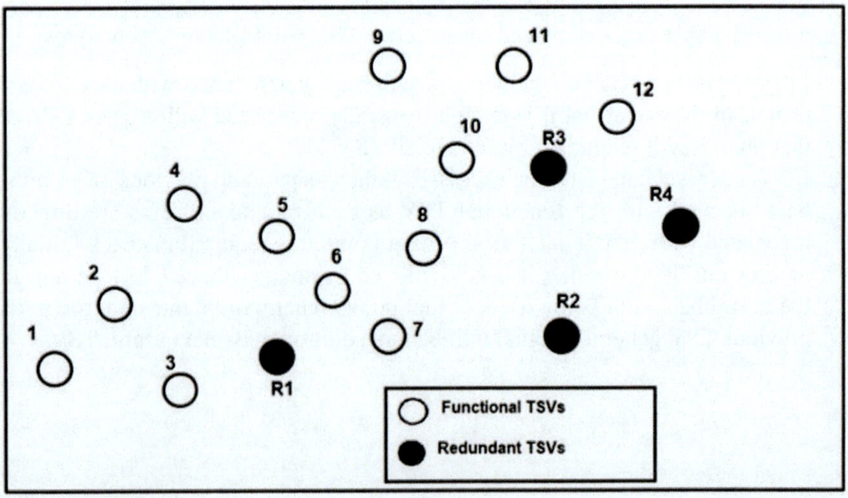

Fig. 2 Number given to functional & redundant TSVs

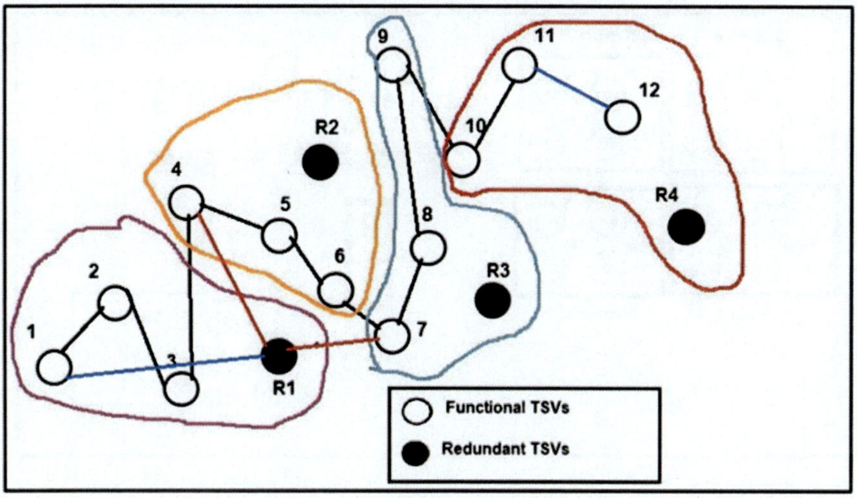

Fig. 3 Grouping and connection of redundant TSVs

and router-based excess TSVs architectures can efficiently repair the clustered faults (compared to the switching or shifting architectures). However, the implementation of these architectures required a large area overhead and a relatively low repair rate when several faults occur in the TSVs. Therefore, our study proposes a new redundant architecture with a high repair rate for multiple clustered faults and a required low area overhead compared to the existing solutions for clustered faults, the redundant architecture proposed in the next section, we utilize both switching and shifting architectures to repair multiple TSV faults we have seen that our method required low overhear and can repair clustered faults. In switching technique, a redundant TSV is used to reroute the signal if any faulty TSV exists in the circuit. In Fig. 3 if redundant TSV "1" is found faulty then the redundant TSV R1 will be used to reroute the signal of functional TSV "1". In shifting the signal can be rerouted through its nearest neighbor. If the functional TSV "2" is found faulty then the signal of TSV "2" will be shifted to functional TSV "3" and signal of TSV "3" will be shifted to functional TSV "4" and the redundant TSV R2 will be used to reroute the signal of functional TSV "4". In this case shifting and switching method is used.

To fix F-TSVs (functional TSV) with a great repair rate, each R-TSV (redundant TSV) ought to incorporate an exact number of sources of info. The signal of each R-TSV is resolved as pursues. The contributions of each R-TSV contain two parts: the first incorporates one gathering's signal and the second incorporates the other gathering's signal. Basically, each gathering contains one R-TSV. Thus, each R-TSV of each gathering involves all signal in its gathering as the data sources. For instance, the MUX of R-TSV R-2 incorporates all signal of the B bunch as its information sources (Fig. 4).

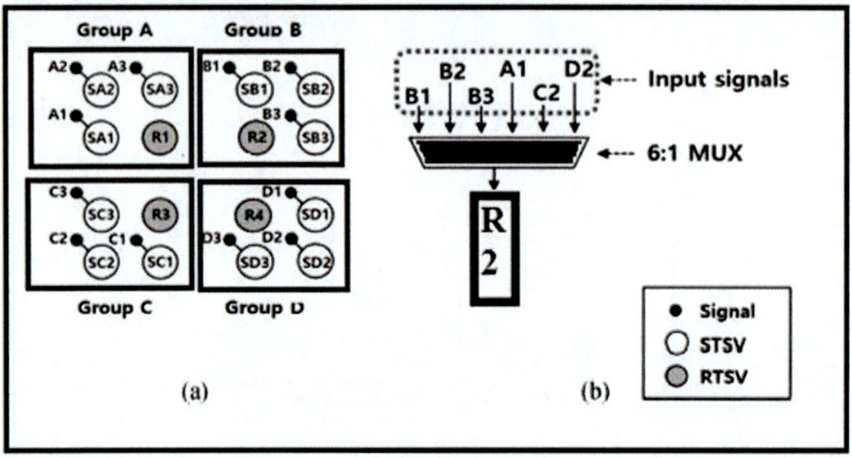

Fig. 4 **a** Diagram of proposed architecture **b** TSV grouping

Likewise, R-5 incorporates all signal of the gathering. The second segment of each R-TSV's signal includes the other gathering's signal. On the off chance that R-2 incorporates all signal of the B gathering, the second segment of R-2's information signal involves just one signal among all signal of different gatherings, with the exception of the B gathering. On the off chance that the quantity of S-TSVs is 56, given us a chance to expect that the number of gatherings that is equivalent to the quantity of R-TSVs is eight and the quantity of individuals in each gathering is seven. Accordingly, in this case, one R-TSV can choose all signal of one gathering and just one sign of the other seven gatherings. Hence, an aggregate of $7 + 7 = 14$ signal can be resending by one R-TSV. To show this R-TSV age process.

Be that as it may, to accomplish a high fix rate for grouped deficiencies, it is not adequate to just redundant F-TSVs with R-TSVs. Allow us to expect that only one R-TSV is unused and only one F-TSV ought to be fixed after the different F-TSVs are fixed. Luckily, if the signal of this unused R-TSV incorporate a F-TSV signal, it tends to be effectively fixed by redundant TSVs; generally, the R-TSV cannot fix the F-TSV. To tackle this kind of issue, the TSVs in each gathering are associated with 2:1 MUXs as a chain, as appeared in Fig. 5. This chain structure encourages the fix of hard-to- switch shortcomings. Furthermore, this is the way to effectively fix grouped blames by creating different rerouting ways if there should be an occurrence of TSV disappointment. The fix procedure is depicted in next segment (Fig. 6).

Repairing Method:

Nonetheless, to accomplish a high fix rate for grouped shortcomings, it is not adequate to just redundant F-TSVs with R-TSVs. Allow us to accept the proposed functional engineering receives two sorts of fix techniques: "exchanging" and "move and-exchanging." These fix techniques are depicted in this area with certain models,

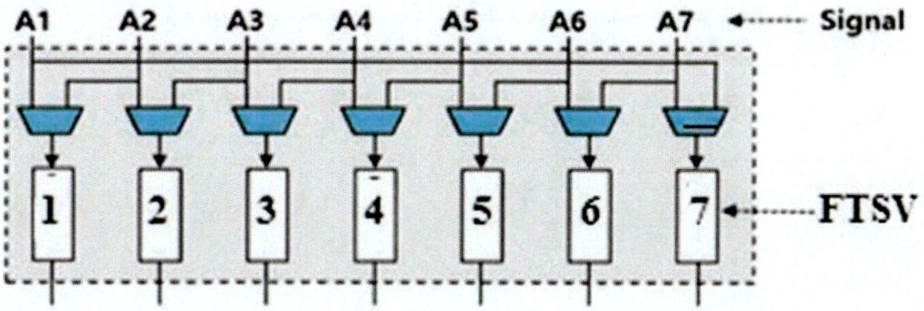

Fig. 5 Multiplexor chain

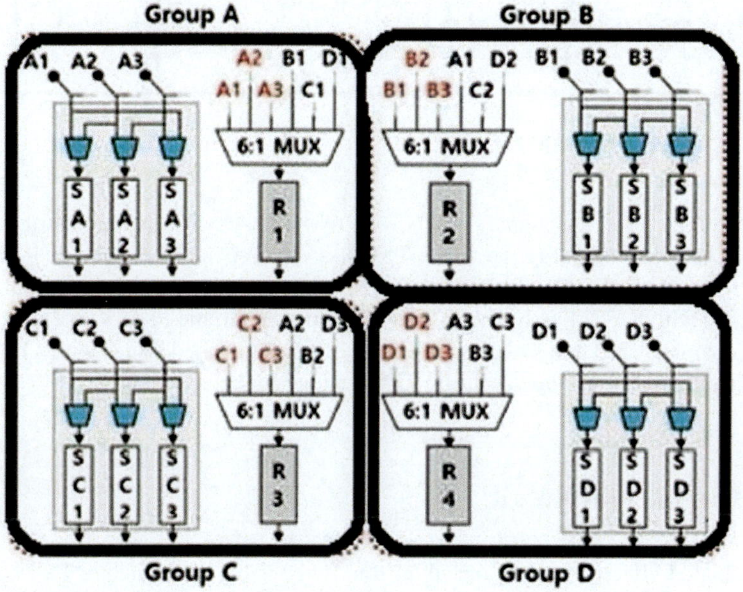

Fig. 6 Proposed architecture with 12:4 ratio

as appeared in Fig. 7, where three number of S-TSVs are associated with 2:1 multiplexer and only one R-TSV is accessible. The TSVs appeared in Fig. 7a incorporate three S-TSVs, with SA-1 being defective. Without a F-TSV in this gathering, the A-1 sign is allocated to SA-1. In any case, if a flaw happens in SA-1, this defective SA-1 can be fixed essentially by rerouting the A-1 sign to the accessible R- TSVs. Within the sight of any R-TSV that can associate the A-1 signal, a MUX that is a piece of the R-TSV chooses the A-1 signal and this R-TSV goes about as the SA-1. This kind of fix component is named "exchanging" in this paper. Be that as it may, without a R-TSV that can dole out signal of the F-TSVs, an elective fix component ought to be connected. Without a R- TSV to relegate the A-3 signal, as appeared in Fig. 7b, the "exchanging" strategy cannot fix SA-3 on the grounds that the R-TSV R-2 alone

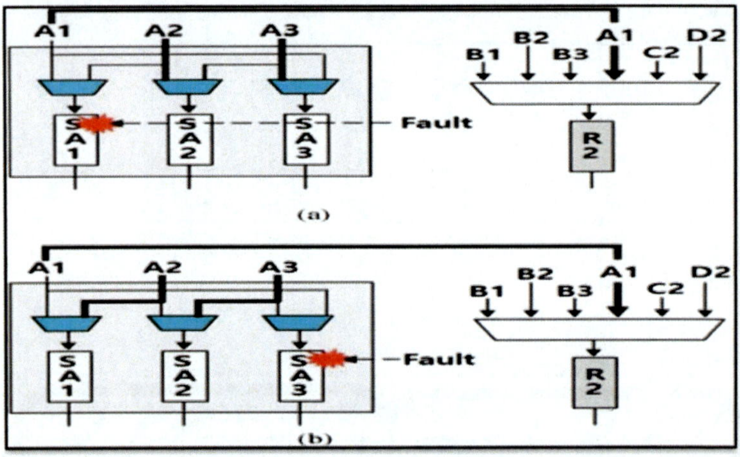

Fig. 7 Repairing of Faulty TSVs. **a** switching. **b** shifting and switching

cannot interface the A-3 signal. In this way, to fix SA-3, a MUX chain structure is utilized. As appeared in Fig. 7b, the R-TSV R-2 can associate the A-1 signal. In this manner, so as to fix SA-3, allocating A-2 sign to SA-1 empowers the A-3 signal to be associated with SA-2. Lastly, the A-1 sign is appointed to the R-TSV R-2; this kind of fix is named "shift and switching." To fix different bunched issues with a high fix rate, it is imperative to apply a fitting fix technique for each deficiency.

6 Experimental Result

We evaluated effectiveness of our algorithm by considering 32 active TSVs and 4 extra TSVs. Table 1 describes the requirement of MUXs and the dependency of various groups. Group number 1, 2, 3 are created by taking the active TSVs between two consecutives excess TSVs. Other groups are created by overlapping these groups.

Table 1 Different groups size with 32 functional and 4 redundant TSVs

Group No	No. of functional TSVs	No. redundant TSVs	No. of require MUXs	Dependency
1	10	2	22	20
2	7	2	16	14
3	12	2	26	24
4	16	3	36	48
5	18	3	40	54
6	27	4	60	108

Table 2 Repair rate compared to others works (Unit-100%)

	64 TSV				36 TSV			
#no of Fault	In [8]	In [13]	In [9]	Proposed	In [8]	In [9]	In [13]	Proposed
1	100	100	100	100	100	100	100	100
2	100	100	100	100	100	100	100	100
3	100	100	100	100	100	99.5	98.6	100
4	99.5	98.5	99.5	100	86.7	96.5	97.8	100
5	97.5	90	98.5	100	85.6	94.3	95	100

Table 2 shows the required MUXs is minimum in our architecture compared to [1, 9, 10, 13]. Though required MUXs of [1] are closed to our work, but in [1] they have taken minimum number of MUXs. If the position changes the number of required MUXs will increase for same number of active and excess TSVs. But our work required number of MUXs are same irrespective of the position of active and excess TSVs. Table 2 shows repair rate with compared to others works.

Tables show that our proposed method can repair equal number of faults of redundant TSVs.

7 Conclusion and Future Scope

We have described an architecture of replacing a faulty TSVs by redundant TSVs. The proposed method of making interconnection between all active TSVs and makes direct association among utilitarian and excess TSVs to create gatherings. We increment the absolute reliance with the goal that we can fix most extreme number of defective TSVs. Likewise, in our proposed technique required number of MUXs is less contrasted with the other existing strategies. Thus, all out territory and equipment cost of the chip will be less. In this work, we have not considered the required wire length to connect functional and redundant TSVs. So in our future work, we will be working to reduce the wire length.

References

1. J. Burun L, Mcllrath C, Keast C, lwes A, Loomis K, Warner, Wyatt P (2001) Three dimensional Integrated circuit for low power, High- Bandwidth system-on-chips. proc of IEEE International Solid-State Circuit Conference (ISSCC), pp 268–269
2. Weerasekera R et al (2007) Extending systems-on chip to the third dimension: performance, cost and technological tradeoffs. Int Conference on Computer-Aided design, pp212–219
3. Chen H, Shih J-Y, Li S-W, Lin H-C, Wang M-J, Peng C-N (2010) Electrical tests for three-dimensional ICs (3dics) with TSVs, In Proc. of 3D Test Workshop Informal Digest

4. Zhao Y, Khursheed S, Al-Hashimi BM (2011) Cost-effective TSV grouping for yield improvement of 3D-ICs. in Proc. Asian Test Symposium, pp 201–206
5. Hsieh A-C, Hwang TT, Chang M-T, Tseng C-M, Li H-C (2010) TSV redundancy: architecture and design issues in 3D IC, pp 166–171
6. Roy SK, Roy K, Giri C, Rahaman H (2015) Recovery of faulty TSVs in 3D ICs. In Proc of 16th Int'l Symposium on Quality Electronics Design, pp 533–536
7. Roy SK, Chatterjee S, Giri C, Rahaman H (2013) Repairing of faulty TSVs using available number of multiplexers in 3D ICs. In Proc. of Asian Symposium on Quality Electronics design (ASQED 2013), pp 155–160
8. Lo W-H, Chi K, Hwang TT (2015) Architecture of ring-based redundant TSV for clustered faults. Design, Automation and Test in Europe Conference and Exhibition (DATE), pp 549–553
9. Jiang L, Xu Q, Eklow B (2012) "On effective TSV repair for 3D-stacked ICs", Design. Automation and Test in Europe Conference and Exhibition (DATE) 2012:793–798
10. Noia B, Chakraborty K (2012) Pre-bond Probing of TSVs in 3D stacked ICs. In Proc. IEEE International test Conference, pp 1–10, September 20–22.
11. Roy SK, Chatterjee S, Giri C (2012) Identifying faulty TSVs in 3D stacked IC during pre-bond testing. In Proc. IEEE International Symposium on electronics System Design (ISED), pp 162–166
12. Kang U, Chung H-J, Heo S (2010) 8 Gb 3-D DDR3 DRAM using through-silicon-via technology. IEEE J Solid-State Circuits 45(1), 111–119
13. Ghosh S, Roy SK, Rahaman H, Giri C (2017) TSV repairing for 3D ICs using redundant TSV". In Proc. 7th IEEE International Symposium on electronics System Design (ISED-2017). pp 1–5

Chapter 5
Study on Similarity Measures in Group Decision-Making Based on Signless Laplacian Energy of an Intuitionistic Fuzzy Graph

Obbu Ramesh, S. Sharief Basha, and Raja Das

Abstract In sight of intuitionistic fuzzy inclination relations (IFIR), we study group decision-making (GDM) problems. We propose another way to deal with assess the relative notoriety weights of specialists by registering the questionable proof of intuitionistic fuzzy inclination relations and the normal similitude level of one individual intuitionistic inclination connection to the others. This new approach takes both objective and subjective evidence of specialists into consideration. Then, we assimilate the weights of authorities into the precise intuitionistic fuzzy inclination relations and progress a relative similarity method to originate the significances of substitutes and better of the substitutes. The balance investigation with extra techniques by two numerical examples shows the sober mindedness and supportiveness of the anticipated strategies.

Keywords Group decision-making · Intuitionistic fuzzy inclination relation · Intuitionistic fuzzy graph · Signless Laplacian energy · Similarity measure

1 Introduction

The intuitionistic fuzzy sets (IFSs) [1], considered by two quite functions like membership and non-membership functions, are commonsense in numerous fields, for example, clinical diagnosis [2], decision making [3, 4] and example acknowledgment [5, 6]. Szimidt and Kacprzyk [7–9] investigated in bunch of decision-making issues dependent on IFSs, the extent of course of action in a gathering of specialists (characters) as the intuitionistic fuzzy inclination relations which are portrayed from singular inclinations. They gave a strategy to total the aggregate fuzzy intuitionistic inclination connection gathered from individual intuitionistic fuzzy preference relations, while the positions may not be framed to substitutes [10]. We realize that

O. Ramesh
Vellore Institute of Technology, Vellore, Tamil Nadu 632014, India

S. S. Basha · R. Das (✉)
Department of Mathematics, Vellore Institute of Technology, Vellore, Tamil Nadu 632014, India

J. K. Mandal et al. (eds.), *Proceedings of International Conference on Innovations in Software Architecture and Computational Systems*, Studies in Autonomic, Data-driven and Industrial Computing, https://doi.org/10.1007/978-981-16-4301-9_5

in Refs. [7–9] given the intuitionistic fuzzy preference relations include of three kinds of matrices. Xu [11] combined one network from these three sorts of networks (matrices) and anticipated the impression of an 'intuitionistic preference connection.' He built up an intuitionistic inclination proof that way to deal with intuitionistic preference relations collective choice creation (group decision-making) that are utilized to amassed by a number-crunching averaging administrator of intuitionistic fuzzy graph and a weighted math averaging administrator of intuitionistic fuzzy diagram.

On GDM issues among the majority of those written works, by power's area fields, and so forth, we perceived the specialists' societal position and ability, which are foreordained the significance weights of specialists and generally saw as the given boundaries. In this article, the 'abstract' weights of specialists are referenced from such sort of given weights. At that time, it is sensibly rousing to speak the way to begin the weights of specialists, referenced to because the objective weights of experts during this paper, from their adjusting intuitionistic inclination relations, which depict the specialists' inclination proof about each pair of substitutes. How to affect evaluate the objective weights of specialists are concentrated during this article by utilizing some proof measure instruments—Signless Laplacian energy and similitude proportions of IFSs. As two significant topics within the hypothesis of fuzzy sets, both Signless Laplacian energy and similarity measures of IFGs are inspected broadly from various view focuses.

The graph of energy was introduced during the 1970's, and from that point forward has been highly concentrated in Laplacian energy and Signless Laplacian energy. To rough approximation the level of likeness (similarity) between two IFSs, we utilized the similarity measures of IFSs. A similarity measure or likeness work is a genuine esteemed capacity that evaluates the similitude between two objects. Although no single meaning of a similarity measure exists, normally such measures are in some sense the backward of separation measurements: They take on huge qualities for comparable objects and either zero or a negative an incentive for exceptionally divergent items.

The remaining of this paper is prearranged as follows. Section 2 shows that the Signless Laplacian energy formulas of IFGs laid out in Refs. [12, 13] separately and also presents an efficient similarity measure for IFGs by an alteration of Signless Laplacian energy measures into similarity measures. Section 3 presents a way to use their subjective and objective weights to work out the weights of authorities and recommends a relative similarity method to rank the substitutes. One example on group deciding problems is shown for instance the efficiency and reasonability of the projected methods by assessments with the others. Section 4 stretches the conclusions.

2 Preliminaries

2.1 Intuitionistic Fuzzy Signless Laplacian Energy

Definition 2.1.1 [See 12]: An intuitionistic fuzzy graph (IFG) is defined as $G =$ (V,E,μ,γ), where V is the set of vertices and E is the set of edges,μ is a fuzzy membership function defined based on V × V, and γ is a fuzzy non-membership function. We define $\mu(v_i, v_j)$ by μ_{ij} and $\gamma(v_i, v_j)$ by γ_{ij} such that (i) $0 \leq \mu_{ij}+\gamma_{ij} \leq 1$ (ii) $0 \leq \mu_{ij,}\gamma_{ij}, \pi_{ij} \leq 1$ where $\pi_{ij} = 1 - \mu_{ij} - \gamma_{ij}$. Hence, $(V \times V, \mu, \gamma)$ is said to be an Intuitionistic fuzzy graph (Fig. 1).

Example 2.1.2 See Fig. 1

Example 2.1.3 For Fig. 1, the adjacency matrix is defined as $A(\tilde{G}) = [(\mu_{ij}), (\gamma_{ij})]$, where μ_{ij} and γ_{ij} are membership non-membership entries.

$$A(\tilde{G}) = \begin{bmatrix} (0,0) & (0.4,0.5) & (0.1,0.8) & (0.5,0.3) \\ (0.4,0.5) & (0,0) & (0.3,0.6) & (0.6,0.1) \\ (0.1,0.8) & (0.3,0.6) & (0,0) & (0.7,0.2) \\ (0.5,0.3) & (0.6,0.1) & (0.7,0.2) & (0,0) \end{bmatrix}$$

$$A(\mu_{ij}) = \begin{bmatrix} 0 & 0.4 & 0.1 & 0.5 \\ 0.4 & 0 & 0.3 & 0.6 \\ 0.1 & 0.3 & 0 & 0.7 \\ 0.5 & 0.6 & 0.7 & 0 \end{bmatrix} \text{ and } A(\mu_{ij}) = \begin{bmatrix} 0 & 0.5 & 0.8 & 0.3 \\ 0.5 & 0 & 0.6 & 0.1 \\ 0.8 & 0.6 & 0 & 0.2 \\ 0.3 & 0.1 & 0.2 & 0 \end{bmatrix}.$$

Definition 2.1.4 Let A($\tilde{G}$) be a adjacency matrix of IFG and (X, Y) are Eigen values of A($\tilde{G}$), where X and Y are Eigen values of corresponding matrices A(μ_{ij}) and A(γ_{ij}).

Definition 2.1.4 Let A($\tilde{G}$) be a adjacency matrix and D($\tilde{G}$) = [d_{ij}] be a degree matrix of IFG $\tilde{G}$ = (V,E,μ, γ). The matrix $L^+(\tilde{G})$ = D($\tilde{G}$) + A($\tilde{G}$) is defined as Signless Laplacian matrix of $\tilde{G}$.

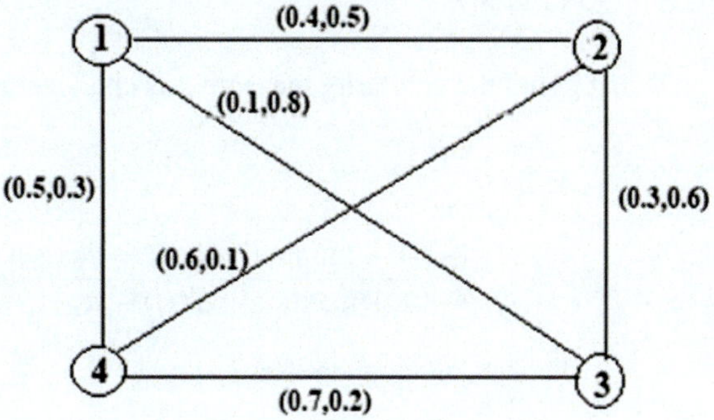

Fig. 1 An intutionistic fuzzy graph

The form of the Signless Laplacian matrix is
$L^{+}(\tilde{G}) = [(L^{+}(\mu_{ij})), (L^{+}(\gamma_{ij}))]$.

Example 2.1.5 Fig. 1 having the following membership and non-membership values of Signless Laplacian matrix of an IFG.

$$L^{+}[\mu_{ij}] = \begin{bmatrix} 1.0\ 0.4\ 0.1\ 0.5 \\ 0.4\ 1.3\ 0.3\ 0.6 \\ 0.1\ 0.3\ 1.1\ 0.7 \\ 0.5\ 0.6\ 0.7\ 1.8 \end{bmatrix} \text{ and } L^{+}[\mu_{ij}] = \begin{bmatrix} 1.6\ 0.5\ 0.8\ 0.3 \\ 0.5\ 1.2\ 0.6\ 0.1 \\ 0.8\ 0.6\ 1.6\ 0.2 \\ 0.3\ 0.1\ 0.2\ 0.6 \end{bmatrix}.$$

Definition 2.1.6 Let $L[\mu_{ij}]]$ be the fuzzy Laplacian matrix IFG of $\tilde{G} = (\text{V,E},\mu, \gamma)$, then the characteristic polynomial of its Laplacian matrix is $\varphi(IG, \mu) = \det(\mu I_n - L(G))$. The roots of $\varphi(IG, \mu)$ are the intuitionistic fuzzy Laplacian Eigen values of IG.

2.2 *Similarity Measures for Intuitionistic Fuzzy Sets (IFS_s)*

Definition 2.2.1 [5, 14]: Let us consider $F(x)$ be a intuitionistic fuzzy set, then normal interval-valued function $\zeta : F(\text{x}) \times F(X) \rightarrow [0, 1][0, 1]$ is called a similarity measure, if ζ satisfies the following properties $\forall A, B, C \in F(X)$:

(i) $0 \leq \zeta(A, B) \leq 1$
(ii) $\zeta(A, B) = 1$ If and only if $A = B$;
(iii) $\zeta(A, B) = \zeta(B, A)$
(iv) If $A \subseteq B \subseteq C$, then $\zeta(A, C) \leq \zeta(A, B)$ and $\zeta(A, C) \leq \zeta(B, C)$.

For A and B in $F(X)$, let

$$u_{M(A,B)}(x_i) = \frac{1 + \text{minima}\{|u_A(x_i) - u_B(x_i)|, |v_A(x_i) - v_A(x_i)|\}}{2} \tag{1}$$

$$v_{M(A,B)}(x_i) = \frac{1 - \text{maxima}\{|u_A(x_i) - u_B(x_i)|, |v_A(x_i) - v_A(x_i)|\}}{2} \tag{2}$$

At that point, the similarity measure ζ is characterized by

$$\zeta(A, B) = \frac{1}{n} \sum_{i=1}^{n} \frac{1 - \text{minima}\{|u_A(x_i) - u_B(x_i)|, |v_A(x_i) - v_B(x_i)|\}}{1 + \text{maxima}\{|u_A(x_i) - u_B(x_i)|, |v_A(x_i) - v_B(x_i)|\}}, \quad for\ A, B \in F(X) \tag{3}$$

Taking into consideration the elements on the globe may have different centrality; here, we depict the weighted kind of condition (formula) (18).

Let the weighting vector $y = (y_1, y_2, y_3, \ldots y_n)$ with $y_i \geq 0$ and $\sum_{i=1}^{n} y_i = 1$ of elements $X_i, \forall i = 1, 2, \ldots n$.

The weighted similarity measure is characterized as

$$\zeta(A, B) = \sum_{i=1}^{n} y_i \frac{1 - \text{minima}\{|u_A(x_i) - u_B(x_i)|, |v_A(x_i) - v_B(x_i)|\}}{1 + \text{maxima}\{|u_A(x_i) - u_B(x_i)|, |v_A(x_i) - v_B(x_i)|\}} \tag{4}$$

Here, the similarity measures and Signless Energy of IFGs are utilized to select the weights of specialists and rank the substitutes for crew decision-making problems reliant on intuitionistic tendency relations.

3 Intuitionistic Inclination Relations

All through the dynamic procedure, an authority is conventionally important to offer his/her tendencies over the substitutes. The authority may give his/her choices in light of a specific objective while from time to time he/she isn't actually certain about those choices. Thusly, it is fitting to convey the position's tendency qualities with intuitionistic fuzzy characteristics instead of the numerical characteristics. 'Szmidt' and 'Kacprzyk' first summarized the fuzzy tendency connection to the intuitionistic fuzzy tendency association containing three kinds of lattices. From that point, by joining the three kinds of matrices into one, 'Xu' presented the probability of an intuitionistic inclination association.

Definition An intuitionistic preference relation 'P' on the set X is spoken to by a matrix $P = \left[p_{ij}\right]_{n \times n}$, where $p_{ij} = \left\langle (x_i, x_j), u(x_i, x_j), v(x_i, x_j) \right\rangle, \quad \forall i, j = 1, 2, \ldots, n$.

Let $p_{ij} = (u_{ij}, v_{ij})$, where p_{ij} is an intuitionistic fuzzy value, composed by the certainty degree u_{ij} to which X_i is preferred to X_j and the certainty degree v_{ij} to which X_i is non-preferred to X_j and $\pi_{ij} = 1 - u_{ij} - v_{ij}$ is interpreted as the uncertainty degree to which X_i is preferred to X_j Moreover u_{ij}, v_{ij} satisfy $0 \leq u_{ij} + v_{ij} \leq 1, \quad \forall i, j = 1, 2, \ldots, n$.

Next, we look at the best approach to get progressively proof from the specialists inclinations over the substitutes to change the given criticalness weights of specialists for intelligently sensible dynamic.

3.1 A Technique to Discover the Weights of Specialists

The organization decision-making trouble taken into consideration on this paper may be portrayed as follows:

Let $X = \{X_1, X_2, \ldots, X_n\}$ be the arrangement of substitutes, $E = \{e_1, e_2, \ldots, e_n\}$ be the arrangement of experts. The authority e_k gives his/her inclination proof for each pair of substitutes and develops an intuitionistic fuzzy inclination connection.

$P^{(k)} = \left(p_{ij}^{(k)}\right)_{n\times n}$, Where $p_{ij}^{(k)} = (u_{ij}^{(k)}, v_{ij}^{(k)}),\quad 0 \leq u_{ij}^{(k)} + v_{ij}^{(k)} \leq 1,\quad \forall i, j = 1, 2, 3, \ldots, n.$

For the GDM hassle supported upheld intuitionistic tendency family members, the mixing of the character intuitionistic tendency family members into an combination intuitionistic tendency connection is anticipated. The comparative significance loads of specialists got the chance to be fused into every individual intuitionistic tendency connection and influence the amassing result.

The weights concerning professionals are related to their convivial positions yet prominence.

Aptitudes saw for clear zones and numerous others yet are often destined for a group dynamic issue. Regardless, the experts choices, for instance, intuitionistic inclination relations made during the basic reasoning procedure may not by and large be considered even those new confirmation reflects their real data on the substitutes. As the weights may accept an overall activity toward the last situating of the substitutes, by then how to give out reasonable loads toward the authorities during the rational powerful system is an issue. Here, the predefined loads of the specialists' significance are viewed as one sort of 'dynamic' weight of the specialists. Separated and the predefined loads, the intuitionistic propensity relations who express the geniuses' tendency affirmation may mirror their confirmed understandings toward the substitutes in a powerfully 'objective' sense, by then the loads of specialists got from their differentiating intuitionistic inclination relations are suggested as the 'objective' heaps of authorities. How to get the sensible 'objective' heaps of specialists? Next, we propose a way to deal with oversee audit the 'objective' loads of specialists utilizing entropy and likeness proportions of IFGs.

The Signless Laplacian energy is able to calculate the unsure evidence of IFG. Every intuitionistic fuzzy inclination relation $P^{(k)}, \quad \forall K = 1, 2, \ldots, n$ is in fact an IFS in $X \times X$; thus, we can quantify its dubious proof by utilizing the Signless Laplacian vitality measure. Consequently, we will quantify its dubious proof by utilizing the Signless Laplacian vitality measure. During the dynamic procedure, we when in doubt expect the weakness level of the intuitionistic tendency relations as meager as attainable for more affirmation of the practiced results. Hence, the more prominent the entropy of $P^{(k)}$ and e_k.Then again, the similarity degree $\zeta(P^{(k)}, P^{(l)})$ among two intuitionistic inclination relations $P^{(k)}$ and $P^{(l)}$ can be estimated by recipe. Next, the average similarity degree of $P^{(k)}$ to the others can be determined; the larger worth, the larger weight given to the authority e_k. As shown by the above assessment, we develop the succeeding Method-I to overview the 'objective' loads of the experts.

To illuminate the ongoing decision-making issues, let us consider subjective weighting vector $y = (y_1, y_2, y_3, \ldots, y_n)$ with $y_i \geq 0$ and $\sum_{i=1}^{n} y_i = 1$ of elements $X_i, \forall i = 1, 2, \ldots, n$.

Step 1. Compute the Signless Laplacian energy $LE^+(P^{(k)})$ of $P^{(k)}$:

$$LE^+(P^{(k)}) = \left| \tilde{\mu}_i - \frac{2 \sum_{1 \leq i \leq j \leq n} \mu(u_i, u_j)}{n} \right| \tag{5}$$

Step 2. Compute the weight y_k^a, resolute by $LE^+(P^k)$ of the authority e_k:

$$\begin{aligned} Y_k &= \left((Y_\mu)_l, (Y_\gamma)_l \right) \\ &= \left[\frac{LE^+((D_\mu)_l)}{\sum_{k=1}^{m} LE^+((D_\mu)_k)}, \frac{LE^+((D_\gamma)_l)}{\sum_{k=1}^{m} LE^+((D_\gamma)_k)} \right] for\, l = 1, 2, \ldots, m \end{aligned} \tag{6}$$

Step 3. Compute $\zeta(P^k, P^l)$ among P^k & P^l for all $k \neq l$:

$$\zeta(P^k, P^l) = \frac{1}{n} + \frac{2}{n^2} \sum_{i=1}^{n} \sum_{j=i+1}^{n} \frac{1 - \text{minima}\{\left|u_{ij}^{(k)} - u_{ij}^{(l)}\right|, \left|v_{ij}^{(k)} - v_{ij}^{(l)}\right|\}}{1 + \text{maxima}\{\left|u_{ij}^{(k)} - u_{ij}^{(l)}\right|, \left|v_{ij}^{(k)} - v_{ij}^{(l)}\right|\}} \tag{7}$$

And, the average degree for similarity $\zeta(P^k)$ of P^k is determined by

$$\zeta(P^k) = \frac{1}{m-1} \sum_{l=1,\, l \neq k}^{m} \zeta(P^k, P^l), \quad \forall k = 1, 2, \ldots, m. \tag{8}$$

Step 4. Conclude the weight y_k^b of authority e_k:

$$y_k^b = \frac{\zeta(P^k)}{\sum_{i=1}^{n} \zeta(P^i)}, \quad \forall k = 1, 2, \ldots, n. \tag{9}$$

Step 5. Work out the weight of objective y_k^2 of authority e_k:

$$y_k^2 = \eta y_k^a + (1 - \eta) y_k^b, \quad \eta \in [0, 1], \quad \forall k = 1, 2, \ldots, m. \tag{10}$$

Step 6. Identify with the weight of subjective (y_k^1) and the weight of objective (y_k^2) into the weight Y_k of authority e_k:

$$y_k = \gamma y_k^a + (1-\gamma) y_k^2, \quad \gamma \in [0, 1], \quad \forall k = 1, 2, 3, \ldots, m. \tag{11}$$

From Method-I, the weights of authorities consider both the subjective and objective verification. We at that point absorb the individual intuitionistic inclination relations into an aggregate intuitionistic preference connection by utilizing the succeeding hypothesis specified by Ref. [13].

Theorem 3 [15]: Let

$$P^{(k)} = [p_{ij}^{(k)}]_{n \times n} \tag{12}$$

be a intuitionistic fuzzy inclination relations specified by the authorities e_k, and $y = (y_1, y_2, y_3, \ldots\ldots y_n)$ be a weighting vector of authorities, where $p_{ij}^{(k)} = (u_{ij}^{(k)}, v_{ij}^{(k)})$, with $y_i \geq 0$ & $\sum_{i=1}^{n} y_i = 1$. Then, the aggregation $P = [p_{ij}]_{n \times n}$ of $P^{(k)} = [p_{ij}^{(k)}]_{n \times n}$ is also an intuitionistic inclination relation, where $p_{ij} = [u_{ij}, v_{ij}]; u_{ij} = \sum_{k=1}^{m} y_k u_{ij}^{(k)}, v_{ij} = \sum_{k=1}^{m} y_k v_{ij}^{(k)}, \forall i, j = 1, 2 \ldots\ldots n$. [15].

Step 7

$$p_i^{(k)} = \frac{1}{n} \sum_{j=1}^{n} p_{ij}^{(k)}, i = 1, 2, 3, \ldots n \tag{13}$$

Formula to obtain the $p_i^{(k)}$ of the substitute x_i over all the other substitutes.

Step 8

$$p_i = \sum_{k=1}^{m} y_k p_i^{(k)} \tag{14}$$

Formula to aggregate all $p_i^{(k)}$, consequent to p authorities, into a combined intuitionistic fuzzy value $p_i(u_i, v_i)$ of the substitute x_i on the whole other substitutes.

Step 9

$$S(m_i) = (u_i - v_i) \tag{15}$$

Compute the score function of m_i and the larger the value of $\zeta(p_i)$, the improved the substitute x_i. Then, the rank of the substitutes is acquired.

3.2 Comparative Similarity Technique Toward Rank the Substitutes

The ith row vector $\{(u_{ij}, v_{ij}),\ \forall i, j = 1, 2, \ldots, n\}$ of a co-operative intuitionistic fuzzy inclination relation M, represented by M^i, defines the pair wise comparison inclination of the ith substitute x_i on the whole the substitutes in X and can be observed seeing that an IFS in $\{x_i\} \times X$. Let X^+ and X^- be the positive perfect and negative perfect substitute, correspondingly. Expect the IFS $R^+ = \{(1, 0), (1, 0), (1, 0), \ldots, (1, 0)\}\ \&R^- = \{(1, 0), (1, 0), (1, 0), \ldots (1, 0)\}$ designate pair wise assessment inclination of X^+ and X^- overall the substitutes in X, respectively.

In this manner, the best substitute is created to be full of the level of comparability to x^+ as large as plausible and contain the level of closeness to x^- as little as likely. Consequently, we are able to rank the substitutes as of the co-employable inclination connection by utilizing the succeeding relative similitude strategy: Method-II. Suppose that $P^{(K)},\ \forall k = 1, 2, \ldots n$ and we. are well précised as earlier.

Method-II

Step 1. Find out co-operative fuzzy inclination relation $P = [p_{ij}]_{n\times n}$ by

$$p_{ij} = (u_{ij}, v_{ij}) = \left(\sum_{k=1}^{m} y_k u_{ij}^{(k)}, \sum_{k=1}^{m} y_k v_{ij}^{(k)}\right), \quad \forall i, j = 1, 2, 3, 4 \ldots, n \tag{16}$$

Step 2. In support of every substitute X_i, find out similarity measure $\zeta(P^i, P^+)$ between P^i and P^+ similarity measure $\zeta(P^i, P^-)$ between P^i and P^- by recipe (11).

$$\begin{aligned}\zeta(P^i, P^+) &= \frac{1}{n}\sum_{j=1}^{n} \frac{1 - \text{minima}\{|u_{ij} - 1|, |v_{ij} - 0|\}}{1 + \text{maxima}\{|u_{ij} - 1|, |v_{ij} - 0|\}} \\ &= \frac{1}{n}\sum_{j=1}^{n} \frac{1 - \text{minima}\{1 - u_{ij}, v_{ij}\}}{1 + \text{maxima}\{1 - u_{ij}, v_{ij}\}}\end{aligned} \tag{17}$$

$$\begin{aligned}\zeta(P^i, P^-) &= \frac{1}{n}\sum_{j=1}^{n} \frac{1 - \text{minima}\{|u_{ij} - 0|, |v_{ij} - 1|\}}{1 + \text{maxima}\{|u_{ij} - 0|, |v_{ij} - 1|\}} \\ &= \frac{1}{n}\sum_{j=1}^{n} \frac{1 - \text{minima}\{u_{ij}, 1 - v_{ij}\}}{1 + \text{maxima}\{u_{ij}, 1 - v_{ij}\}}\end{aligned} \tag{18}$$

Step 3. For every substitute X_i, find out its calculation value

$$g(x_i) = \frac{\zeta(P^i, P^+)}{\zeta(P^i, P^+) + \zeta(P^i, P^+)} \tag{19}$$

The more prominent estimation of $f(x_i)$, the prevalent the substitute x_i. At that point, the position of the substitutes is acclimatized. The succeeding two examples are given to tell the best way to accomplish the acclimatized weights through Method-I and how to rank the substitutes with Method-II.

3.3 *Illustrations*

Now, we demonstrate two illustrations by tolerant one available illustration taken by Xu and Yager [13] and an additional one by Gong et al. [16]. Through appraisal by way of strategies in Refs. [15] and [16], we attempt to explain these techniques portrayal more proof which is not uncovered previously.

Illustration 1. Let us consider four substitutes $X_i, \quad \forall i = 1, 2, 3,$ and 4 and three authorities $e_k, \quad \forall k = 1, 2$ and 3 in decision-making issue. Assume the weights for every authority are 0.5, 0.3 and 0.2, correspondingly. By comparing the four substitutes, we can form the fuzzy inclination relation for every authority $e_k, \quad \forall k = 1, 2$ and 3.

$P^{(k)} = [p_{ij}^k]_{4\times 4}, \forall k = 1, 2, 3$ and 4 Shown as follows

$$P^{(1)} = \begin{bmatrix} (0,0) & (0.1,0.6) & (0.2,0.4) & (0.7,0.3) \\ (0.5,0.2) & (0,0) & (0.3,0.6) & (0.4,0.2) \\ (0.3,0.5) & (0.4,0.3) & (0,0) & (0.6,0.2) \\ (0.2,0.7) & (0.5,0.4) & (0.2,0.8) & (0,0) \end{bmatrix}$$

$$P^{(2)} = \begin{bmatrix} (0,0) & (0.3,0.5) & (0.1,0.5) & (0.4,0.3) \\ (0.2,0.3) & (0,0) & (0.4,0.3) & (0.5,0.3) \\ (0.5,0.2) & (0.4,0.1) & (0,0) & (0.8,0.2) \\ (0.1,0.6) & (0.3,0.1) & (0.2,0.5) & (0,0) \end{bmatrix}$$

$$P^{(3)} = \begin{bmatrix} (0,0) & (0.3,0.1) & (0.3,0.5) & (0.1,0.4) \\ (0.1,0.5) & (0,0) & (0.5,0.3) & (0.4,0.5) \\ (0.5,0.3) & (0.3,0.2) & (0,0) & (0.3,0.6) \\ (0.4,0.3) & (0.6,0.4) & (0.3,0.3) & (0,0) \end{bmatrix}$$

Example 1 was executed by Xu and Yager [15] for understanding investigation in decision-making dependent on fuzzy tendency relations, however, for inclining components (diagonal values) of the contiguousness matrices are shown as zeros as indicated by fuzzy charts which shows no association (connection). Here, we utilize the information to start the positioning request of the substitutes.

The three authorities observed the original subjective weights as 0.5, 0.3, 0.2, respectively. Then, the subjective weighting vector is $y_1 = (0.5, 0.3, 0.2)$. By using objective and subjective weighting vectors, we can find assimilated weighting vectors of the authorities by Method-1.

From Eq. (5), the Signless Laplacian energies of $P^{(i)}, \forall i = 1, 2,$ and 3.

$LE^{+}(P^{1}) = (4.5050,\ 5.2166)$, $LE^{+}(P^{2}) = (4.2000, 3.9000)$, $LE^{+}(P^{3}) = (4.1000, 4.4435)$.

By Eq. (6), we acquire the weighting vector of the specialists controlled by the Signless Laplacian energies.

$y_1^a = (0.3518,\ 0.3847)$, $y_2^a = (0.3280,\ 0.2876)$, $y_3^a = (0.3202,\ 0.3277)$.

Using Eq. (7), we obtain.

$$\zeta(P^{(1)}, P^{(2)}) = 0.8352,$$

$$\zeta(P^{(1)}, P^{(3)}) = 0.7091,$$

$$\zeta(P^{(2)}, P^{(3)}) = 0.7874.$$

Next, using Eqs. (8) and (9), the averaged similarity degrees $\zeta(p^{(i)})$ of $p^{(i)}, \quad \forall i = 1,\ 2,$ and 3 and the weighting vector $y^{(b)}$ of the authorities determined by the average similarity degrees.

$\zeta(p^{(1)}) = 0.7722$, $\zeta(p^{(2)}) = 0.8113$, $\zeta(p^{(3)}) = 0.7578$

$$y_1^b = 0.3312,\ y_2^b = 0.3479,\ y_2^b = 0.3209$$

Let us take $\eta = 0.5$, which measures sponsors half to the goal.

By solving Eq. (10), we obtain the objective weighting vectors.

$$y_1^2 = (0.3415, 0.3580),\ y_2^2 = (0.3380, 0.3178),\ y_3^2 = (0.3206, 0.3243)$$

From Eq. (11), the assimilated weighting vector 'y' can be calculated by using the subjective weighting vector y^a and the objective weighting vector y^2. That is$y_k = \gamma y_k^a + (1-\gamma)y_k^2, \quad \gamma \in [0, 1]$, where γ is resolute by the decision-makers rendering to their inclinations to the 'objective' and 'subjective' weight indication. First, we assume $\gamma = 0.5$ in Eq. (11). We get the assimilated weighting vector y:

$$y_1 = (0.3467, 0.3714);\ y_2 = (0.3330, 0.3027);\ y_3 = (0.3204,\ 0.3260)$$

Up to this, we found the assimilated weights of authorities for the sensible decision-making issue.

Initially, we depict Xu's disposition to build up the decision result, which contains the succeeding advances:

Step 1 Formula:$p_i^{(k)} = \frac{1}{(n-1)} \sum_{j=1}^{n} p_{ij}^{(k)}, \forall i = 1, 2, 3, \ldots, n$ to get the averaged intuitionistic fuzzy value $p_i^{(k)}$ of the substitute x_i over all the other substitutes:

$$p_1^{(1)} = (0.3333,\ 0.4333)\quad p_2^{(1)} = (0.4000,\ 0.3333).$$
$$p_3^{(1)} = (0.4333,\ 0.3333)\quad p_4^{(1)} = (0.3000,\ 0.6333).$$

$$p_1^{(2)} = (0.2667,\ 0.4333)\quad p_2^{(2)} = (0.3667,\ 0.3000).$$
$$p_3^{(2)} = (0.5667,\ 0.1667)\quad p_4^{(2)} = (0.2000,\ 0.4000).$$

$$p_1^{(3)} = (0.2333,\ 0.3333)\quad p_2^{(3)} = (0.3333,\ 0.4333).$$
$$p_2^{(3)} = (0.3667,\ 0.3667)\quad p_4^{(3)} = (0.4333,\ 0.3333).$$

Step 2 Formula $p_i = \sum_{k=1}^{n} y_k p_i^{(k)}, \quad \forall i = 1, 2, \ldots, n$
to aggregate all $p_i^{(k)}$,(k = 1, 2,…,m), corresponding to m authorities, into a collective intuitionistic fuzzy value $p_i(u_i, v_i)$ of the substitute $\mathrm{x_i}$ over all the other substitutes:

$p_1 = (0.2790, 0.3681)$, $p_2 = (0.3676, 0.3559)$,$p_3 = (0.4564, 0.2905)$, $p_4 = (0.3094, 0.4649)$.

Step 3 Compute the score function $S(m_i) = u_i - v_i$ of m_i and get.
$\zeta(p_1) = -0.0891$, $\zeta(p_2) = 0.0117$, $\zeta(p_3) = 0.1659$, $\zeta(p_4) = -0.1555$.
Then,$\zeta(p_3) > \zeta(p_2) > \zeta(p_1) > \zeta(p_4)$ and.
Hence,$x_3 > x_2 > x_1 > x_4$.
By comparative similarity method, we obtain the ranking order.

Procedure II

We obtain the collective fuzzy inclination relation by Eq. (16) in Method-II,

$$P = \begin{bmatrix} (0,0) & (0.2307, 0.4068) & (0.1988, 0.4629) & (0.4079, 0.3326) \\ (0.2720, 0.3281) & (0,0) & (0.3974, 0.4115) & (0.4333, 0.3281) \\ (0.4307, 0.3440) & (0.3680, 0.2069) & (0,0) & (0.5705, 0.3304) \\ (0.2308, 0.5394) & (0.4655, 0.3092) & (0.2321, 0.5463) & (0,0) \end{bmatrix}$$

By Eq. (17), we obtain.

$$\zeta(P^1, P^+) = 0.3882 \quad \zeta(P^2, P^+) = 0.4213$$
$$\zeta(P^3, P^+) = 0.4681 \quad \zeta(P^4, P^+) = 0.3668$$

By formula (18), we obtain

$$\zeta(P^1, P^-) = 0.4648 \quad \zeta(P^2, P^-) = 0.4135$$
$$\zeta(P^3, P^-) = 0.3634 \quad \zeta(P^4, P^-) = 0.4395$$

Then, Eq. (19) gives the evaluation values of substitute's $x_i (i = 1, 2, 3 \text{ and } 4)$:
$g(x_1) = 0.4551$,$g(x_2) = 0.5047$,$g(x_3) = 0.5630$,$g(x_4) = 0.4104$.

Since $g(x_3) > (x_2) > (x_1) > (x_4)$.

We have,$x_3 > x_2 > x_1 > x_4$.

By Method-II, we can say that x_3 ranks the topmost, x_4 ranks the last, however, x_2 and x_1 have middle position order. The ranking orders are same in both the methods.

Section 3.4 gives the calculation outcome by Xu's loom for dissimilar weighting vectors of authority consequent to dissimilar γ values, and Sect. 3.5 inclines the result for Method-II.

By equating the position outcome as listed in both Sects. 3.4 and 3.5, we catch that Xu's loom [11] and method-II give the equal position orders for $\gamma = 0.2$ and $\gamma = 0$. On the other hand, for different values $\gamma = 1$, $\gamma = 0.8$, $\gamma = 0.5$, $\gamma = 0.2$ and $\gamma = 0$, the ranking orders in both methods are same. The results expected by Method-II are distorted close by the divergent segments of objective and subjective parts inside the all-out weight of a power. We may perceive that the positioning result contrast with unique weights among the specialists. The results of Method-II duplicate the impacts of the deviations of the weights.

3.4 Verifying the Ranking Order for Different Γ values by Method-I

Case 1: When $\gamma = 1$, we have

$$y_1 = (0.3518,\ 0.3847);\ y_2 = (0.3280,\ 0.2876);\ y_3 = (0.3202,\ 0.3277)$$

And $\begin{matrix} p_1 = (0.2794,\ 0.4049); & p_2 = (0.3676,\ 0.3563) \\ p_3 = (0.4560,\ 0.2953); & p_4 = (0.3097,\ 0.4667) \end{matrix}$.

Then, $\begin{matrix} \zeta(p_1) = -0.1255 & \zeta(p_2) = 0.0112 \\ \zeta(p_3) = 0.1594 & \zeta(p_4) = -0.1580 \end{matrix}$.

The ranking order is: $x_3 > x_2 > x_1 > x_4$.

Case 2: When $\gamma = 0.8$, we have

$$y_1 = (0.3497,\ 0.3794);\ y_2 = (0.3300,\ 0.2937);\ y_3 = (0.3203,\ 0.3270)$$

And $\begin{matrix} p_1 = (0.2793,\ 0.3627); & p_2 = (0.3676,\ 0.3563) \\ p_3 = (0.4560,\ 0.2953); & p_4 = (0.3097,\ 0.4667) \end{matrix}$.

Then, $\begin{matrix} \zeta(p_1) = -0.0834; & \zeta(p_2) = 0.0113 \\ \zeta(p_3) = 0.1607; & \zeta(p_4) = -0.1570 \end{matrix}$.

The ranking order is: $x_3 > x_2 > x_1 > x_4$.

Case 3: When $\gamma = 0.5$, we have.

$$y_1 = (0.3467,\ 0.3714);\ y_2 = (0.3330,\ 0.3027);\ y_3 = (0.3204,\ 0.3260)$$

And $\begin{array}{ll} p_1 = (0.2790,\ 0.3681); & p_2 = (0.3676,\ 0.3559) \\ p_3 = (0.4564,\ 0.2905); & p_4 = (0.3094,\ 0.4649) \end{array}.$

Then, $\begin{array}{ll} \zeta(p_1) = -0.1255; & \zeta(p_2) = 0.0112 \\ \zeta(p_3) = 0.1594; & \zeta(p_4) = -0.1580 \end{array}.$

The ranking order is: $x_3 > x_2 > x_1 > x_4$.

Case 4: When $\gamma = 0.2$, we have

$$y_1 = (0.3436,\ 0.3633);\ y_2 = (0.3360,\ 0.3118);\ y_3 = (0.3205,\ 0.3250)$$

And $\begin{array}{ll} p_1 = (0.2789,\ 0.4008); & p_2 = (0.3675,\ 0.355) \\ p_3 = (0.4568,\ 0.2922); & p_4 = (0.3092,\ 0.4631) \end{array}.$

Then, $\begin{array}{ll} \zeta(p_1) = -0.1219; & \zeta(p_2) = 0.0120 \\ \zeta(p_3) = 0.1646; & \zeta(p_4) = -0.1539 \end{array}.$

The ranking order is: $x_3 > x_2 > x_1 > x_4$.

Case 5: When $\gamma = 0$, we have

$$y_1 == (0.3415,\ 0.3580);\ y_2 = (0.3380,\ 0.3178);\ y_3 = (0.3206,\ 0.3243)$$

And $\begin{array}{ll} p_1 = (0.2788,\ 0.4009); & p_2 = (0.3674,\ 0.3552) \\ p_3 = (0.4571,\ 0.2912); & p_4 = (0.3090,\ 0.4619) \end{array}.$

Then, $\begin{array}{ll} \zeta(p_1) = -0.1221; & \zeta(p_2) = 0.0122 \\ \zeta(p_3) = 0.1659; & \zeta(p_4) = -0.1529 \end{array}.$

The ranking order is: $x_3 > x_2 > x_1 > x_4$.

We noticed that the ranking order is similar for all the above five cases by Method-I.

3.5 *Verifying the Ranking Order for Different Γ Values by Method-II*

Case 1: When $\gamma = 1$, we have

$$y_1 = (0.3518,\ 0.3847);\ y_2 = (0.3280,\ 0.2876);\ y_3 = (0.3202,\ 0.3277)$$

And $\begin{array}{ll} \zeta(P^1, P^+) = 0.3883; & \zeta(P^2, P^+) = 0.4209; \\ \zeta(P^3, P^+) = 0.3655; & \zeta(P^4, P^+) = 0.4524; \end{array}.$

Now, $\begin{array}{ll} \zeta(P^1, P^-) = 0.4646; & \zeta(P^2, P^-) = 0.4135; \\ \zeta(P^3, P^-) = 0.4022; & \zeta(P^3, P^-) = 0.4682 \end{array}.$

Then, $g(x_1) = 0.4553,\ g(x_2) = 0.5044,\ g(x_3) = 0.5294,\ g(x_4) = 0.4384$.

The ranking order is: $g(x_3) > (x_2) > (x_1) > (x_4)$.

Hence, $x_3 > x_2 > x_1 > x_4$.

Case 2: When $\gamma = 0.8$, we have

$$y_1 = (0.3518,\ 0.3847);\ y_2 = (0.3280,\ 0.2876);\ y_3 = (0.3202,\ 0.3277)$$

And $\begin{matrix} \zeta(P^1, P^+) = 0.3883; & \zeta(P^2, P^+) = 0.4211; \\ \zeta(P^3, P^+) = 0.4673; & \zeta(P^4, P^+) = 0.3660; \end{matrix}$.

Now, $\begin{matrix} \zeta(P^1, P^-) = 0.4647; & \zeta(P^2, P^-) = 0.4135; \\ \zeta(P^3, P^-) = 0.3638; & \zeta(P^4, P^-) = 0.4680 \end{matrix}$.

Then, $g(x_1) = 0.4552,\ g(x_2) = 0.5046,\ g(x_3) = 0.5623,\ g(x_4) = 0.4388$.
The ranking order is: $g(x_3) > (x_2) > (x_1) > (x_4)$.
Hence, $x_3 > x_2 > x_1 > x_4$.

Case 3: When $\gamma = 0.5$, we have

$$y_1 = (0.3518,\ 0.3847);\ y_2 = (0.3280,\ 0.2876);\ y_3 = (0.3202,\ 0.3277)$$

And $\begin{matrix} \zeta(P^1, P^+) = 0.3882; & \zeta(P^2, P^+) = 0.4213; \\ \zeta(P^3, P^+) = 0.4681; & \zeta(P^4, P^+) = 0.3668; \end{matrix}$

Now, $\begin{matrix} \zeta(P^1, P^-) = 0.4648; & \zeta(P^2, P^-) = 0.4135; \\ \zeta(P^3, P^-) = 0.3634; & \zeta(P^4, P^-) = 0.4395 \end{matrix}$.

Then, $g(x_1) = 0.4551,\ g(x_2) = 0.5047,\ g(x_3) = 0.5630,\ g(x_4) = 0.4140$.
The ranking order is: $g(x_3) > (x_2) > (x_1) > (x_4)$.
Hence, $x_3 > x_2 > x_1 > x_4$.

Case 4: When $\gamma = 0.2$, we have

$$y_1 = (0.3518,\ 0.3847);\ y_2 = (0.3280,\ 0.2876);\ y_3 = (0.3202,\ 0.3277)$$

And $\begin{matrix} \zeta(P^1, P^+) = 0.3881; & \zeta(P^2, P^+) = 0.4216; \\ \zeta(P^3, P^+) = 0.4689; & \zeta(P^4, P^+) = 0.3676; \end{matrix}$

Now, $\begin{matrix} \zeta(P^1, P^-) = 0.4649; & \zeta(P^2, P^-) = 0.4134; \\ \zeta(P^3, P^-) = 0.3630; & \zeta(P^4, P^-) = 0.4675 \end{matrix}$.

Then, $g(x_1) = 0.4550,\ g(x_2) = 0.5049,\ g(x_3) = 0.5636,\ g(x_4) = 0.4402$.
The ranking order is: $g(x_3) > (x_2) > (x_1) > (x_4)$.
Hence, $x_3 > x_2 > x_1 > x_4$.

Case 5: When $\gamma = 0$, we have

$$y_1 = (0.3518,\ 0.3847);\ y_2 = (0.3280,\ 0.2876);\ y_3 = (0.3202,\ 0.3277)$$

And $\begin{array}{l}\zeta(P^1, P^+) = 0.3880; \quad \zeta(P^2, P^+) = 0.4216; \\ \zeta(P^3, P^+) = 0.4695; \quad \zeta(P^4, P^+) = 0.3681;\end{array}$

Now, $\begin{array}{l}\zeta(P^1, P^-) = 0.4650; \quad \zeta(P^2, P^-) = 0.4134; \\ \zeta(P^3, P^-) = 0.3627; \quad \zeta(P^4, P^-) = 0.4673\end{array}.$

Then, $g(x_1) = 0.4549,\ g(x_2) = 0.5049,\ g(x_3) = 0.5642,\ g(x_4) = 0.4406.$

The ranking order is: $g(x_3) > (x_2) > (x_1) > (x_4)$.

Hence, $x_3 > x_2 > x_1 > x_4$.

We noticed that the ranking order is similar for all the above five cases by Method-II.

Hence, we obtained the similar results for Method-I and Method-II.

4 Conclusion

Newly, numerous entropy and similarity measures formulae are signed to the GDM issues made on intuitionistic fuzzy proof. In this paper, we apply the similarity measures and Signless Laplacian vitality (energy) on the uncertain proof of intuitionistic fuzzy diagram inclination relations and the normal similitude level of one explicit intuitionistic fuzzy chart preference connection to the others, separately. We build up a technique to gauge the essentialness weights of specialists by considering both the subjective and objective weights of the specialists. The subjective a part of the weights denotes to the normal weights usually predetermined consistent with the authorities' social or academic fame or administrative positions actually. We extract evidence from the authorities' practical judgments (individual intuitionistic inclination relations) toward the substitutes and transform it into the objective weights of the authorities by Method-I. We also aggregate the individual intuitionistic predilection relations into a combined intuitionistic inclination relation by using an intuitionistic fuzzy weighted arithmetic averaging operator and propose a relative similarity method to derive the priorities of substitutes from the combined intuitionistic predilection relation by Method-II.

References

1. Atanassov K (1986) Intuitionistic fuzzy sets. Fuzzy Sets Syst 20(1):87–96
2. De SK, Biswas R, Roy AR (2001) An application of intuitionistic fuzzy sets in medical diagnosis. Fuzzy Sets and Syst 117(2):209–213
3. Atanassov K, Pasi G, Yager RR (2008) Intuitionistic fuzzy interpretations of multi criteria multi-person and multi-measurement tool decision making. Int J Sys Sci 36:859–868
4. Gong ZW, Li LS, et al (2010) On additive consistent properties of the intuitionistic fuzzy predilection relation. Int J Evidence Tech Decision Mak 9(6):1009–1025.
5. Xu ZS (2007) On similarity measures of interval-valued intuitionistic fuzzy sets and their application to pattern recognitions. J Southeast Univ 23(1):139–143 (in English)

6. Vlochos IK, Sergiadis GD (2007) Intuitionistic fuzzy evidence—applications to pattern recognition. Pattern Recognition Lett 28:197–206
7. Szmidt E, Kacprzyk J (2001) Analysis of consensus under intuitionistic fuzzy predilections. In: Proceedings of the International Conference Fuzzy Logic and Technology, September 5–7, De Montfort Univ. Leicester, UK, pp 79–82
8. Szmidt E, Kacprzyk J (2002) Analysis of agreement in a group of authorities via distances between intuitionistic fuzzy predilections. In: Proceedings of the 9th International Conference IPMU 2002, Annecy, France, 2002, pp 1859–1865
9. Szmidt E, Kacprzyk J (2005) A new concept of a similarity measure for intuitionistic fuzzy sets and its use in group decision making. In: V. Torra, Y. Narukawa, S. Miyamoto (Eds) Modelling decision for artificial intelligence, LNAI 3558,pp 272–282. Springer
10. Szmidt E, Kacprzyk J (2002) Using intuitionistic fuzzy sets in group decision making. Control and Cybernetics **31**, 1037–1053
11. Xu ZS (2007) Intuitionistic predilection relations and their application in group decision making. Evidence Sciences 177(4):2363–2379
12. Parvathi R, Karumbigai M G, Intutionistic fuzzy graphs, computational intelligence. Theory and Appl 139–150.
13. Sharbaf SR, Fayazi F (2014) Laplacian Energy of a fuzzy graph. Iranian J Mathematical Chem 5(1):1–10
14. Xu ZS (2008) An overview of distance and similarity measures of intuitionistic sets. Internat J Uncertain Fuzziness Knowledge-Based Systems 16(4):529–555
15. Xu ZS, Yager RR (2009) Intuitionistic and interval-valued intuitionistic fuzzy predilection relations and their measures of similarity for the evaluation of agreement within a group. Fuzzy Optimizat Decision Making 8(2):123–129
16. Gong ZW, Li LS, Zhou FX, Yao TX (2009) Goal programming approaches to obtain the priority vectors from the intuitionistic fuzzy predilection relations. Computers and Industrial Eng 57(2009):1187–1193
17. Gutman I, Zhou B (2006) Laplacian Energy of a Graph. Linear Algebra Appl 414(1):29–37

Chapter 6
Asymptotic Stability of Neural Network System with Stochastic Perturbations

M. Lakshmi and Raja Das

Abstract In this article, we have considered a mathematical model of a neural network. The model is characterised by a delay difference equation with stochastic perturbations. We have proved the condition of asymptotic stability behaviour of the trivial solution of s multiple-delay model with stochastic terms.

Keywords Asymptotic stability · Lyapunov-Krasovkii functional · Delay difference equation · Stochastic difference equation · Neural network

1 Introduction

Difference equations [1] have number of applications in computer science, signal processing, queuing theory, time series analysis, etc. The finite difference schemes considered to solve differential equations (ODE, PDE) are difference equations [2]. Difference equations with special parameters time delay and stochastic perturbation are more powerful in modelling real life phenomena [3, 4]. The stability analysis of neural network has been discussed from Global asymptotic stability [5]. And also, the stability of deterministic type of delay difference equation involving delay [6] is considered. It contains single isolated neuron involving delay. Here, the study on dynamics of a neural network [6] is extended by including multiple delay and stochastic terms. There exist many results on stability issues of recurrent neural networks (RNNs) either with constant delay [7–13] or with time varying delay [14–17]. The aim of the research is to derive stability conditions for the stochastic delay difference equations [18].

M. Lakshmi · R. Das (✉)
School of Advanced Sciences, VIT, Vellore 632 006, India

J. K. Mandal et al. (eds.), *Proceedings of International Conference on Innovations in Software Architecture and Computational Systems*, Studies in Autonomic, Data-driven and Industrial Computing, https://doi.org/10.1007/978-981-16-4301-9_6

2 Mathematical Model

The following nonlinear multiple-delay differential equation represents a neural network.

$$dy(t) = \left[-y(t) + \alpha\phi\left(y(t) - \sum_{i=1}^{r} \beta_i y(t-\tau_i)\right)\right]dt, \quad t > 0 \tag{1}$$

Fig. 1 represents the Illustration of Multi-delay differential neural network model.
where

$y(t)$ denotes the activation level of a neuron at time t,
$\alpha > 0$ is a real constant relating the range of the continuous variable $y(t)$,
$\beta \geq 0$ denotes the measure of the influence of the past history,
$\tau \in [0, \infty)$ is the time delay, s
$\phi : \mathbb{R} \to \mathbb{R}$is a nonlinear continuous function such that $|\phi(u)| \leq |u| \forall u \in \mathbb{R}$.

Equation (1) is extended by including the stochastic perturbations as given.

$$\mathrm{d}y(t) = \left[-y(t) + \alpha\phi\left(y(t) - \sum_{i=1}^{r} \beta_i y(t-\tau_i)\right)\right]\mathrm{d}t + \xi(t, y(t), y(t-\tau_1), \ldots, y(t-\tau_r))dW_t, \quad t > 0 \tag{2}$$

where $W = \{W(t)\}_{t \geq 0}$ is the one dimensional Wiener process, τ_i, $i = 1, \ldots, r$ are non-random numbers such that $0 < \tau_1 < \ldots < \tau_r = \tau$, $\xi : \mathbb{R}^{k+2} \to \mathbb{R}$ is a nonlinear continuous function.

The Euler–Maruyama discretization of Eq. (2) leads to the following delay difference equation with stochastic perturbations.

$$Y_{k+1} = (1-d)Y_k + \alpha d\phi\left(Y_k - \sum_{i=1}^{r} \beta_i Y_{k-i}\right) + \sqrt{d}\xi(k, Y_k, Y_{k-1}, \ldots, Y_{k-r})\psi_{k+1}, \quad k \in \mathbb{N}_0 \tag{3}$$

with arbitrary non-random initial values $Y_0, Y_{-1}, \ldots Y_{-r} \in \mathbb{R}$. $d \in (0, 1]$ is the step size, ψ_k are independent random variables with $E(\psi_k) = 0$ and $Var(\psi_k) = 1$.

3 Stability of Stochastic Multi-Delay Difference Equation

Consider Eq. (3) and the conditions with $E(\psi_k) = 0$ and $Var(\psi_k) = 1$., where continuous function Ø satisfying the following condition

$$|\emptyset(u)| \le |u|, \quad \forall u \in R \tag{4}$$

Suppose that the function ξ is non-random and there exist $c_i \ge 0, \gamma_k \in R$, $i = 0, 1, \ldots r, \quad n \in N$, such that $\forall u_i \in R, i = 0, 1, \ldots r$, satisfies

$$|\xi(k, u_0, u_1, \ldots u_r)|^2 \le \sum_{i=0}^{r} c_i |u_i|^2 + \gamma_k^2, \quad \xi(k, 0, 0, \ldots, 0) = 0 \tag{5}$$

$$\sum_{i=1}^{\infty} \gamma_i^2 < \infty, \tag{6}$$

$$\alpha^2 (1 + \sum_{i=1}^{r} |\beta_i|)^2 + \sum_{j=0}^{r} c_j < 1 \tag{7}$$

Theorem (2.1) Let Eqs. (4), (5), (6) and (7) be fulfilled. Then, $\lim_{k \to \infty} Y_k = 0$ is asymptotically stable. Where Y_k is a solution of Eq. (6) with arbitrary $d \in (0, 1]$.

To prove this theorem, we require the following lemmas.

Lemma (2.1) Let $\{y_k\}_{k \in N}$ is a sequence of independent $\Im_k$-measurable, $E|y_k| = 0$, $E|y_k| < \infty$. Let also $\{f_k\}_{k \in N}$ is a sequence $\Im_k$-measurable such a way that $E|f_{k-1} y_k| < \infty \, \forall \, k \ge 1$. Then $\{z_k\}_{k \in N}$, $Z_k = \sum_{i=1}^{k} f_{i-1} y_i \; \forall k \in N$, is an $\Im_k$-martingale difference.

Lemma (2.2) If $\{\psi_k\}_{k \in N}$ be an $\Im_k$-martingale difference. Then, there exists an $\Im_k$-martingale difference $\{\mu_k\}_{k \in N}$ as well as positive $\Im_{k-1}$-measurable stochastic sequence $\{\eta_k\}_{k \in N}$ such that for every hen $k \in N$,

$$\psi_k^2 = \mu_k + \eta_k, \quad a.s.$$

If ψ_k is an independent for all $n \ge 0$, then $\eta_k = E(\psi_k^2)$, $\mu_k = \psi_k^2 - E(\psi_k^2)$.

Lemma (2.3) Let $\{Z_k\}_{k \in N}$ is a sequence of non-negative $\Im_k$-measurable, $E|Z_k| < \infty \, \forall \, k \in N$ and

$$Z_{k+1} < Z_k + u_k - v_k + v_{k+1}, \quad k \in N_0$$

where $\{v_k\}_{k \in N}$ is an $\Im_k$-martingale difference, $\{u_k\}_{k \in N}$, $\{v_k\}_{k \in N}$ are non-negative $\Im_k$-measurable process, $E|u_k|$, $E|v_k| < \infty \; \forall \; k \in N$. Then

$$\left\{\omega : \sum_{k=1}^{\infty} u_k < \infty\right\} \subseteq \left\{\omega : \sum_{k=1}^{\infty} v_k < \infty\right\} \cap \{Z_k \to\}$$

By $\{Z_k \rightarrow\}$, we denote the set of all $\omega \in \Omega$ for which $\lim_{k\to\infty} Z_k(\omega)$ exist and is finite.

Proof of the Theorem (2.1). By squaring Eq. (3), we have

$$Y_{k+1}^2 = \left[(1-d)Y_k + \alpha d\varphi\left(Y_k - \sum_{i=1}^{r} \beta_i Y_{k-i}\right)\right]^2 + 2\left[(1-d)Y_k + \alpha d\varphi\left(Y_k - \sum_{i=1}^{r} \beta_i Y_{k-i}\right)\right]\left[\sqrt{d}\xi(k, Y_{k-r}, \ldots Y_k)\psi_{k+1}\right] + \left[d\xi^2(k, Y_{k-r}, \ldots .Y_k)\right]\psi_{k+1}^2 \tag{8}$$

Here, $\{\rho_k\}_{k\in N}$ is an $\Im_k$-martingale difference is defined by

$$\rho_{k+1} = 2\left[(1-d)Y_k + \alpha d\varphi\left(Y_k - \sum_{i=1}^{r} \beta_i Y_{k-i}\right)\right] \left[\sqrt{d}\xi(k, Y_{k-r} \ldots Y_k)\psi_{k+1} + d\xi^2(k, Y_{k-r,} \ldots Y_k)\left[\psi_{k+1}^2 - 1\right]\right] \tag{9}$$

By assumption, we have $E\psi_{k+1}^2 = 1$ and $\left\{\psi_{k+1}^2 - 1\right\}_{k\in N}$ is $\Im_{k+1}$-martingale difference. Then, using Lemma (4), it is clear that both terms on the R.H.S of Eq. (9) are $\Im_{k+1}$-martingale difference. By Eq. (4), we estimate that

$$\begin{aligned}\left|(1-d)Y_k + \alpha d\varphi\left[Y_k - \sum_{i=1}^{r} \beta_i Y_{k-i}\right]\right| &\le |(1-d)|Y_k + |\alpha|d\left|\varphi\left(Y_k - \sum_{i=1}^{r} \beta_i Y_{k-i}\right)\right| \\ &\le |(1-d)|`Y_k + |\alpha|d\left|\varphi\left(Y_k - \sum_{i=1}^{r} \beta_i Y_{k-i}\right)\right| \\ &\le (1-d)|Y_k + |\alpha||d\left[|Y_k| + \sum_{i=1}^{r} |\beta_i||Y_{k-i}|\right] \\ &\le (1-d)|Y_k| + |\alpha|d|Y_k| + |\alpha|d\sum_{i=1} |\beta_i||Y_{k-i}| \\ &\le (1-d+|\alpha|d)|Y_k| + |\alpha|d\sum_{i=1}^{r} |\beta_i||Y_{k-i}|\end{aligned}$$

Let $x_1 = 1 - d + |\alpha|d$, $Y_1 = |Y_k|$, $x_i = |\alpha|d|\beta_{r-1}|$, $Y_i = |Y_{k+1-i}|$ where $i = 2$, 3,..., r + 1 and applying the Holder's inequality, we have

$$\left[(1-d+|\alpha|d)|Y_k|+|\alpha|d\sum_{i=1}^{r}|\beta_i||Y_{k-i}|\right]^2 \leq \left[1-d+|\alpha|d+|\alpha|d\sum_{i=1}^{r}|\beta_i|\right]$$
$$\left[(1-d+|\alpha|dY_k^2+|\alpha|d\sum_{i=1}^{r}|\beta_i|\,Y_{k-i}^{2}\right] \tag{10}$$

(10) and (5), we estimate the R.H.S of Eq. (8)

$$\begin{aligned}
Y_{k+1}^{2} \leq &\left[(1-d+|\alpha|d)|Y_k|+|\alpha|d\sum_{i=1}^{r}|\beta_i|\right]\\
&\left[(1-d+|\alpha|d)|Y_k|+|\alpha|d\sum_{i=1}^{r}|\beta_i|\,Y_{k-i}^{2}\right]+d\sum_{i=1}^{r}c_i\,y_{k-i}^{2}+d\gamma_k^2+\rho_{k+1}\\
\leq &\left[\left((1-d+|\alpha|d)|Y_k|+|\alpha|d\sum_{i=1}^{r}|\beta_i|\right)(1-d)+|\alpha|d+dc_0\right]Y_k^2\\
&+d\sum_{i=1}^{r}\left[|\alpha|\left(1-d+|\alpha|d+|\alpha|d\sum_{j=1}^{r}|\beta_j|\right)|\beta_j|+c_i\right]Y_{k-i}^{2}+d\gamma_k^2+\rho_{k+1}
\end{aligned} \tag{11}$$

Let $\widetilde{c_i}=0\,for\,i=-r,-r+1,\ldots,0,\ r+1,\ r+2,\ldots$ and let

$$\widetilde{c_i}=|\alpha|\left(1-d+|\alpha|d\sum_{i=1}^{r}|\beta_i|\right)|\beta_i|+c_i$$

For $i=1, 2, \ldots,$ r for $k\in N$, we define

$$V_k^{(2)}=d\sum_{i=-r}^{k-1}{Y_i}^2+\sum_{i=-r}^{k-1}\widetilde{c_j},\quad V_k=Y_k^2+V_k^{(2)} \tag{12}$$

To find the increment of $V_k^{(2)}$:

$$\begin{aligned}
\Delta V_k^2&=d\sum_{i=-r}^{k}Y_k^2\sum_{j=k+1-i}^{\infty}\widetilde{c_j}-d\sum_{i=-r}^{k-1}Y_i^2\sum_{j=k-i}^{\infty}\widetilde{c_j}\\
&=d\sum_{i=-r}^{k-1}Y_i^2\sum_{j=k+1-i}^{\infty}\widetilde{c_j}+dY_k^2\sum_{j=1}^{\infty}\widetilde{c_j}-d\sum_{i=-r}^{k-1}Y_i^2\sum_{j=k-i}^{\infty}\widetilde{c_j}
\end{aligned}$$

$$= d\sum_{i=-r}^{k-1} Y_i^2 \sum_{j=k-i}^{\infty} \widetilde{c_j} - d\sum_{i=-r}^{k-1} Y_i^2 \widetilde{c_{k-i}} + dY_k^2 \sum_{j=1}^{\infty} \widetilde{c_j} - d\sum_{i=-r}^{k-1} Y_i^2 \sum_{j=k-i}^{\infty} \widetilde{c_j}$$

$$= d\sum_{i=-r}^{k-1} Y_i^2 \widetilde{c_{k-i}} + dY_k^2 \sum_{j=1}^{\infty} \widetilde{c_j}$$

Since $\widetilde{c_i} = 0 fori \geq r+1$, then the term $d\sum_{i=-r}^{k-1} Y_i^2 \widetilde{c_{k-i}}$ will be as follows.

$$d\sum_{i=-r}^{k-1} Y_i^2 \widetilde{c_{k-i}} = d\sum_{i=1}^{r} Y_i^2 \widetilde{c_{k-i}}$$

$$\Delta V_k^2 = -d\sum_{i=1}^{r} Y_{k-i}^{2} \widetilde{c_i} - dY_k^2 \sum_{j=1}^{\infty} \widetilde{c_j} \tag{13}$$

Applying (11) and (13), we get

$$\begin{aligned}
\Delta V_k^{(2)} &= Y_{k+1}^2 - Y_k^2 + V_{k+1}^{(2)} - V_k^{(2)} \\
&\leq \Bigg\{\Bigg[\Bigg((1-d+|\alpha|d) + |\alpha|d\sum_{i=1}^{r}|\beta_i|\Bigg)(1-d+|\alpha|d) + dc_0\Bigg] \\
&+ d\sum_{i=1}^{r}\Bigg[|\alpha|\Bigg(1-d+|\alpha|d+|\alpha|d\sum_{j=1}^{r}|\beta_j|\Bigg)|\beta_j| + c_i\Bigg]\Bigg\} \\
&\times Y_k^2 + d\gamma_k^2 + \rho_{k+1} - Y_k^2 \\
&\Bigg\{\Bigg[\Bigg((1-d+|\alpha|d) + |\alpha|d\sum_{i=1}^{r}|\beta_i|\Bigg)\Bigg((1-d+|\alpha|d) + |\alpha|d\sum_{j=1}^{r}|\beta_j|\Bigg)\Bigg] \\
&+ d\sum_{j=0}^{r} c_j\Bigg\} Y_k^2 + d\gamma_k^2 + \rho_{k+1} - Y_k^2 \\
&= \Bigg\{\Bigg(1 - d\Bigg[1 - \Bigg|\alpha\Bigg(1+\sum_{i=1}^{r}|\beta_i|\Bigg)\Bigg|\Bigg]\Bigg)^2 - 1 + d\sum_{j=0}^{r} c_j\Bigg\} Y_k^2 + d\gamma_k^2 + \rho_{k+1}
\end{aligned} \tag{14}$$

From Eq. (7), we obtain that $\forall d \in (0, 1]$,

$$0 < 1 - |\alpha|\Bigg(1+\sum_{i=1}^{r}|\beta_j|\Bigg) < 1, \quad 0 < h\Bigg[1 - |\alpha|\Bigg(1+\sum_{i=1}^{r}|\beta_j|\Bigg)\Bigg] < 1$$

$$0 < 1 - h\left[1 - |\alpha|\left(1 + \sum_{i=1}^{r}|\beta_j|\right)\right] \le |\alpha|\left(1 + \sum_{i=1}^{r}|\beta_i|\right)$$

And also $\left(1 - d\left[1 - \alpha\left(1 + \sum_{i=1}^{r}\beta_i\right)\right]\right)^2 + h\sum_{j=0}^{k}c_j \le a^2\left(1 + \sum_{i=1}^{r}\beta_i\right)^2 + \sum_{j=1}^{r}c_j < 1$.

We denote $\varepsilon = 1 - \left(1 - d\left[1 - \alpha\left(1 + \sum_{i=1}^{r}|\beta_i|\right)\right]\right)^2 - h\sum_{j=0}^{r}c_j$.

Then, Eq. (14) takes the form

$$V_{k+1} \le V_k - \varepsilon Y_k^2 + dY_k^2 + \rho_{k+1} \tag{15}$$

Let $Z_k = V_k, u_k = d\gamma_k^2, V_k = \varepsilon Y_k^2 and V_{k+1} = \rho_{k+1}$ and applying the lemma (6) to the inequality (15). Since $P\left\{\omega : \sum_{k=1}^{\infty}u_k < \infty\right\} = P\left\{\omega : \sum_{k=1}^{\infty}V_k < \infty\right\} = 1$ by the condition (6), both the limits $\lim_{k\to\infty}Z_k = \lim_{k\to\infty}V_k$ and $\lim_{k\to\infty}\sum_{i=1}^{k}V_i = \varepsilon\lim_{k\to\infty}\sum_{i=1}^{k}Y_i{}^2$ exist and are a.s. finite.

This concludes the proof.

4 Conclusion

The model of nonlinear multiple-delay differential equations which represent a neural network is considered. The model is extended with stochastic perturbations in the form of Wiener processes. The method of Euler–Maruyama is used to derive the discrete analogue of the differential model. We observed that the multiple-delay difference equation in the presence of stochastic fluctuations is asymptotically stable. The future scope will be the above multi-delay differential equation neural network model apply to the covid-19 cases according to the time delay.

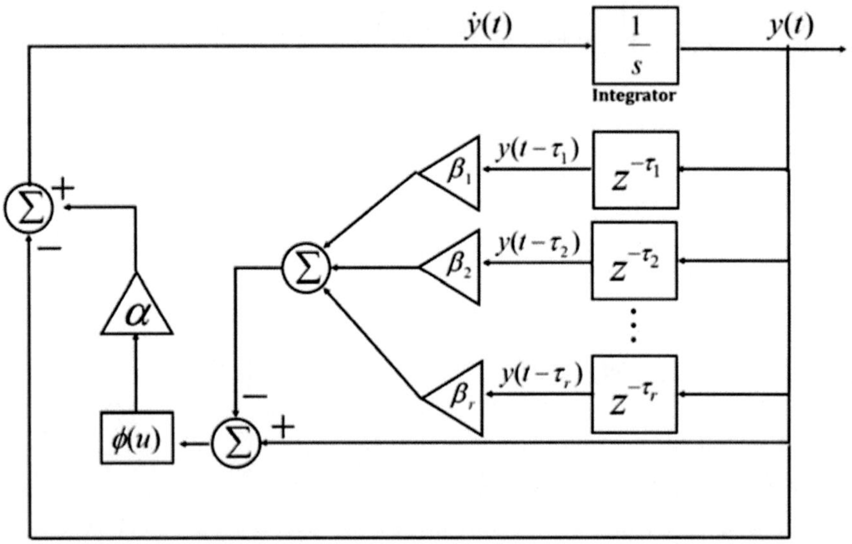

Fig. 1 Neural network model by MATLAB simulink

References

1. Elaydi S (2008) An introduction to difference equations, 3rd edn. Springer International Edition, First Indian Reprint
2. Mickens RE (2015) Difference equations theory, applications and advance topics, 3rd edn. CRC Press, USA
3. Agarwal R P, Wong PJY (1997) Advanced topics in difference equations. Kluwer Academic Publishers
4. Appleby J, Mao X, Rodkina A (2006) On Stochastic stabilization of difference equations. Dynamics of Continuous and Discrete Systems 15(3):843–857
5. Ensari T, Arik S (2005) Global stability analysis of neural networks multiple time varying delays. IEEE Trans Autom Control 50(11):1781–1785
6. Elizabeth S, Nirmal Veena S (2017) Study on asymptotic stabilization of non-linear delay difference equations. Int J Difference Equations 12(1):15–20
7. Ammar B, Cherif F, Alimi AM (2012) Existence and uniqueness of pseudo almost-periodic solutions of recurrent neural Networks wiath time-varying coefficients and mixed delays. IEEE Trans Neural Netw Learn Syst 23(1):109–118
8. Cao J, Zhou D (1998) Stability analysis of delayed cellular neural networks. Neural Netw 51(11):1601–1605
9. Hu J, Wang J (2012) Global stability of complex-valued recurrent neural networks with time-delays. IEEE Trans Neural Netw Learn Syst 23(6):853–865
10. Huang T, Li C, Duan S, Starzyk J (2012) Robust exponential stability of uncertain delayed neural networks with stochastic perturbation and impulse effects. IEEE Trans Neural Netw Learn Syst 23(6):866–875
11. Liu Z, Zhang H, Zhang Q (2010) Novel stability analysis of recurrent neural networks with multiple delays via line integral-type L-K functional. IEEE Trans Neural Netw 21(11):1710–1718
12. Mou S, Gao H, Lam J, Qiang W (2008) A new criterian of delay-dependent asymptotic stability for Hopfield neural networks with time-delay. IEEE Trans Neural Netw 19(3):532–535

13. Stoica A-M, Yaesh I (2008) Markovian jump delayed Hopfield networks with multiplicative noise. Automatica 44(8):2157–2162
14. Liu B, Lu W, Chen T (2011) Generalized Halanay inequalities and their applications to neural networks with unbounded time-varying delays. IEEE Trans Neural Netw 22(9):1508–1513
15. Liu X, Teo KL, Xu B (2005) Exponential stability of impulsive high order Hopfield-type neural networks with time-varying delays. IEEE Trans Neural Netw 16(6):1329–1339
16. Xu S, Lam J (2006) A new approach to exponential stability analysis of neural networks with time-varying delays. Neural Netw 19(1):76–83
17. Yu G-J, Lu C-Y, Tasi JS-H, Su TJ, Liu BD (2003) Stability of cellular neural networks with time-varying delay. IEEE Trans Circuits Syst Part I, 50(5):677–679
18. Rodkina A, Schurz H (2004) A Theorem on asymptotic stability of solutions of non-linear stochastic difference equations with Volterra type noise. SACTA 6(1):23–34

Chapter 7
Uniform Grid Formation by Asynchronous Fat Robots

Moumita Mondal, Sruti Gan Chaudhuri, and Punyasha Chatterjee

Abstract In this paper, we introduce a distributed algorithm for the formation of a grid pattern by multiple homogeneous, autonomous, disc-shaped robots (also referred to as fat robots). The robots organize themselves to form a grid (not given in prior) which is required for covering, safeguarding or supervising a geographical region by the robots. The operation of the robots is based on execution cycles of the phases 'wait–look–compute–move'. In the 'wait' phase, the robots are inactive or dormant; in the 'look' phase, they observe within their visibility range; in the 'compute' phase, the robots decide a destination to move to; and finally, they advance to their decided target point in the 'move' phase. Eventually, the robots distribute themselves to create a grid. They are oblivious, i.e. the robots cannot recollect any past action or looked data from a previous cycle, and there is no interaction among these robots by message passing. They are anonymous, identical and have unlimited visibility. Though the robots are see-through or transparent, they are physical obstruction for the others. The algorithm proposed here also ensures the prevention of deadlock and collision among the robots.

Keywords Grid formation · Mobile fat robots · Asynchronous · Autonomous · Oblivious

1 Introduction

The study on unified behaviour of a troop of tiny self-governing mobile agents is a crucial area of exploration in the domain of robotics. An expensive hefty complex robot may be substituted by a bunch of small autonomous mobile robots also known as *swarm robots* [6] working collaboratively. These swarm robots are essentially movable autonomous programmable particles. The goals of these swarm robots have been patrolling, perceiving and exploring in severe conditions like a disaster-prone area, or in a challenging habitat like under the deep ocean and in the space with

M. Mondal (✉) · S. G. Chaudhuri · P. Chatterjee
Jadavpur University, Kolkata, India

J. K. Mandal et al. (eds.), *Proceedings of International Conference on Innovations in Software Architecture and Computational Systems*, Studies in Autonomic, Data-driven and Industrial Computing, https://doi.org/10.1007/978-981-16-4301-9_8

a requirement for minimal human interference [19, 22]. The decentralized or distributed control of these homogeneous robots with constrained abilities is a popular field of investigation for a long time. Theoretical representation of swarm robots in the Euclidean space has attracted researchers. An important task for executing a job in coordination is to create geometric shapes on the plane by positioning of the robots. This pattern formation task has significantly supported in several fields, like operations in severe environments, military operations, space missions, tumour abscission, etc. [13]. In the current work, we address one of such task of forming a uniform grid, by placing the robots equidistant apart to create a grid pattern in finite time. Each robot collects data of its surrounding environment or neighbouring robots by taking a photograph, then employ this information to compute its destination and finally move to this computed destination. If in the compute phase at a given cycle, no suitable target point could be found, and the robot waits in its current position. A significant application of the algorithm, during COVID-19 pandemic, maybe in covering an area with autonomous robots, scattered in a grid, having UV ray emitting capabilities to disinfect the area [1, 9, 23]. They are oblivious, anonymous, homogeneous and cannot communicate by message passing but sense by observing other robot positions. Considering this robot model, we analyse the problem of grid formation by oblivious fat robots in finite time.

1.1 Framework

A group of fat mobile robots is considered where the radius of these circular robots is assumed to be unit distance. The robots are endowed with (a) movement capabilities, such that they are mobile in the Euclidean space and (b) sensing capabilities, where they observe the position of other robots by taking a photo. Each robot has the ability to look at another robot within its sensing or visibility range (limited or unlimited). Here, we consider unlimited and unrestricted visibility range of the robots. They are anonymous, with no unique identifier and homogeneous, with all executing the same algorithm. As a result of autonomous nature of this robot system, these mobile agents are enabled to perform their task without any centralized control. They are assumed to be free of any kind computational and structural fault. However, every robot has been represented by its centre. Though these robots are transparent in nature, they act as physical hindrance in the path for other robots. The robots undergo rigid movement where they reach their destinations without any interruption. They do not keep any information saved for the future, i.e. oblivious. The robots possess their own local coordinate systems, but they agree on the orientation and direction and of both the axes (i.e. X-axis and Y-axis).
Every robot executes cycles of the following four phases:

- **Wait** The robots are inactive or dormant in this state.
- **look** Here, the robots observe and obtain the positions of all the robots within its range of vision.

- **Compute** Here, the robots execute our algorithm for computing their destinations.
- **Move** In this phase, the robots advance towards their computed targets and finally reach there.

The robots follow one of the following synchronization models:

- In **fully synchronous (FSYNC)** model, every robots operate their cycle at the same time. Thus, the robots fetch the same view in the look phase and compute on the same obtained data.
- In **semi-synchronous (SSYNC)** model, an arbitrary set of robots performs their cycles simultaneously. Here, we have a common clock, but all the robot may or may not be active at each cycle. This kind of scheduling confirms that when any robot is in movement, no other robots is sensing.
- The **asynchronous (ASYNC)** model is a more pragmatic model, where the robots are totally independent. Thus, when a robot finishes its computation in the compute phase, some of the other robots may have already moved from their positions.

Here, we consider the ASYNC model and propose a robot's movement strategy such that, after a definite time, all the robots are settled equidistant apart to form a grid pattern.

2 Our Contribution

In this paper, we prove that formation of a grid pattern by asynchronous fat robots is possible without encountering any collision and deadlock. Starting from a random initial configuration, it ensures uniform coverage by robots in a grid pattern. The aim is to reach a static equilibrium state where the robots are uniformly distributed in a grid. The problem is of significantly important because of its connection with other fundamental robot coordination problems, like scattering, load balancing, exploration, relocation of self-driven electric cars (robots) to recharge stations (nodes), etc. The protocol is decentralized and fully localized. It makes minimal assumptions like; there is no requirement of any direct communication between the robots; they are asynchronous, oblivious, anonymous and identical. Here, we propose a distributed algorithm to form a uniform grid by oblivious asynchronous fat robots with unlimited visibility. One of the main concerns is to avoid collisions among the robots. We have considered transparent robots to avoid obstructed visibility. This transparency can be achieved by enabling the robots to sense their surroundings through ultrasound or a specially designed camera or having a satellite view to get the locations of other robots. Thus, if the robots follow the rigid motion and have agreement on the direction and orientation of both the axes, we show that they are able to create a grid pattern (Fig. 2), without experiencing any deadlock or collision.

3 Earlier Works

A huge volume of research exists in the field of multiple autonomous mobile robots manifesting coordinated behaviour. The primary objective of such a study has been to explore the issues of group architecture, cooperation or collaboration, resource conflict and geometric problems [5]. A newly emerging field of autonomous mobile robots looks at robots as distributed mobile agents or programmable particle and explores various coordination problems with them. They try to solve these problems deterministically supplying correctness proofs for the algorithms. A popular computational model for the robots in this field is called *weak model* [11]. A distributed algorithm specifies which operations must be performed by the robot to achieve the given goal. All the robots execute the same self-deployment algorithm and within finite time, they reach the desired goal. Depending on the assumptions, the existing protocols contrast each other. Some of the primary differences are based on the following criteria:

- Nature of the environment—The robots may move on a Euclidian plane, called continuous (e.g. [4]), or in a graph or network, called discrete (e.g. [2, 14]);
- The robots' actions may be synchronized (e.g. [7, 14]) or without any timing assumptions (e.g. [4, 16]);
- Memory of robots—The robots may have a constant memory (e.g. [7, 13]) or are oblivious (e.g. [4]);
- Visibility or sensing range of robots—This may be limited (e.g. [13, 14]) or unlimited, i.e. cover the entire plane (e.g. [4]);
- Computational power—The robots may be plain finite state machines (e.g. [2, 14]) or may possess computational capability of Turing machines (e.g. [15]);
- Nature of termination of protocol—Exact or approximate covering of the protocol is achieved in definite time (e.g. [4, 14]) otherwise, the protocol converges without reaching a definite or conclusive end (e.g. [7, 10]);
- Type of protocol (generic/specific)—It may be generic, i.e. it works in any system/territory (e.g. [4, 15]) or it is limited for certain groups of graphs/localities (e.g. [7, 10]).

Barriere et al. [3] discussed the problem of uniform scattering for a group of mobile autonomous agents placed on a grid structure where point robots with a constant memory (non-oblivious) and limited visibility are considered. The dispersion problem of mobile robots on graphs is also discussed by Kshemkalyani et al. [17] where they discussed dispersion on a grid graph, ingrained on the Euclidean plane and unfolded solutions that minimize both time to accomplish the dispersion and the memory requirement of each robot. This research on efficient dispersion of robots on a grid/graph is of comprehensive investigation in several fields (e.g. [14, 20, 21]. These papers are on the discrete domain, and the robots considered are also point robots. Czyzowicz et al. [8] had introduced the first algorithmic conclusion on gathering robots characterized by two-dimensional figures (disc shaped robots, also known as fat robots) rather than points in the plain. Pattern formation by autonomous mobile robots on a given plane is one of the most popular problems (e.g. [4, 10, 12, 16, 24]). In [18], an unified self-distribution and coverage maintenance scheme are proposed, which settles the coverage enhancement problem. The uniform covering or self-distribution problem appear essentially when the autonomous mobile robots are deployed arbitrarily in an area, but the requirement is for the area to be uniformly covered with maximum coverage.

4 Algorithm

On the 2D plane, a cluster of n stationary points, representing the centres of the fat robots, is given. Every robot acts as a physical obstruction for the other robots. Since these robots have a radius of unit distance, they have agreement over unit distance. The number of robots, n and the minimum distance to be maintained between the centres of two adjacent robots on the uniform grid (say k, where $k > 3$) is given. Here, we consider $k > 3$, as we are taking fat robots with unit radius and the minimum gap needs to be maintained between the adjacent grid points, while the target positions are filled with the fat robots. The purpose of the algorithm is to determine the robot movements, such that in finite time, the robots are positioned equidistant apart in a grid pattern.

4.1 Underlying Model

Let $R = r_1, r_2, \ldots, r_n$ be a group of autonomous disc-shaped mobile robots, recognized as *fat robots*. The robots are illustrated by their centres, i.e. r_i represents a robot whose centre is r_i. The robots R are characterized as:

- Autonomous.
- Anonymous and homogeneous.
- Oblivious where they cannot recall any action or data from previous cycles.
- The robots have rigid movement.
- They cannot directly communicate by passing messages. Each robot is allowed to have a camera that can take pictures over 360°. They interact only by means of recognizing other robots' positions.
- Fat robots (i.e. the robots have dimensions and here the radius of the robots is considered as unit).
- Transparent robots, but they are hindrance in the path for the other robots.
- They have unlimited visibility.
- Each robot carries out a cycle of wait–look–compute–move asynchronously.
- A robot considers its position as origin (i.e. its local coordinate system). They do not possess any global origin. However, they comply on the orientation and direction of X and Y axes.

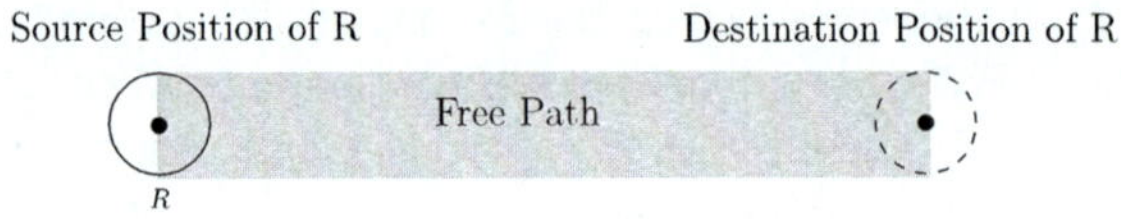

Fig. 1 Free path example

4.2 Overview of the Problem

A cluster of robots R is studied. Our aim is to form a uniform grid of asynchronous oblivious fat robots with unlimited visibility, by moving the robots to the nodes of the grid. The following steps are executed by each robot in its compute phase:

- Determination of grid dimensions, i.e. the number of rows and columns, rc;
- The distance between two adjacent points of the grid is calculated, dist_{adj};
- The target points on the grid are determined using the *ComputeGrid* routine.
- The robots move to the target points of the grid, and a uniform grid of n robots is formed using the *FormGrid* routine.

The robots move in *free paths* to ensure that the movements are free of collision.

Definition 1 Free path is a route of a robot from its source to destination (Fig. 1) such that the rectangular field, with length as the distance bounded by the centres of the source and destination positions and width as two units (since fat robots with radius as unit, consequently diameter of two units), does not contain part of any other robot.

4.3 Description of the Algorithm ComputeGrid

Initially, the robots are scattered on a 2D plane. Let the number of robots in R be n. So, the number of rows and columns of the grid will be $\lceil\sqrt{n}\rceil$ (say rc). Also, let the minimum distance between two adjacent points on the grid, k be an input. To compute the minimum width requirement (say, $grid_{width_{\text{min}}}$) of the grid to be formed, the distance bounded by the robot on the extreme west and the robot on the extreme east is calculated, i.e. the variation between the X-axis value of the west-most robot (say X_{MIN}) and the X-axis value of the east-most robot (say X_{MAX}). Thus, The X-axis value of the extreme west robot (X_{MIN}) is the westward bound of the grid. To determine the X_{MIN}, a robot ri compares the X-axis values of all the robots. So, the robot with the minimum X-axis value is the west-most robot, and its X-axis value is considered as the X_{MIN} (Algorithm 1). Similarly, the X-axis value of the extreme east robot (X_{MAX}) is the eastward bound of the grid. To determine the X_{MAX} also, a robot ri compares the X-axis values of all the robots. The robot with the maximum X-axis value is the east-most robot, and its X-axis value is considered as the X_{MAX} (Algorithm 2). Although the robots locally compute these values, they all agree to the same robot as west-most and thus have the same X_{MIN}, as they have compliance on the sense of orientation and direction of the axes. Similarly, the robots also acknowledge the same robot as the east-most and thus have same X_{MAX}.

So, the possible distance enclosed by the centres of two contiguous robots (say x_{length}) will be $\lfloor grid_{width_{\text{min}}} \div rc \rfloor$. Now, the distance between two adjoining robots on the grid, $dist_{adj}$ is computed by comparing x_{length} and the given input, k. So, $dist_{adj}$ is the maximum of the two lengths, x_{length} and k. Thus making $dist_{adj}$ always exceeding or equal to the given minimum distance k. The width of the final computed grid $grid_{width}$ is computed as $dist_{adj} * rc$ (Fig. 2). Algorithm *ComputeGrid* computes a grid with the equidistant target points on it (Algorithm 4)

Algorithm 1 FindXMIN(n)

Input: n
Output: The west-most bound of the distribution, X_{MIN}
$r_i \in R$ considers its position as origin (0,0).
Let X_{MIN} be X-value of the west-most robot, and initially, $X_{MIN} \leftarrow$ X-value of r_1;
Let c be the robot counter to compare all robots and $c \leftarrow 2$;
while $c <= n$ **do**
    if *X-axis value of* $r_c < X_{MIN}$ **then**
        $X_{MIN} \leftarrow$ X-axis value of r_c;
    $c \leftarrow c + 1$;
return X_{MIN}

Algorithm 2 FindXMAX(n)

Input: n
Output: The east-most bound of the distribution, X_{MAX}
$r_i \in R$ considers its position as origin (0,0).
Let X_{MAX} be X-value of the east-most robot, and initially, $X_{MAX} \leftarrow$ X-value of r_1;
Let c be the robot counter to compare all robots and $c \leftarrow 2$;
while $c <= n$ **do**
    if *X-axis value of* $r_c > X_{MAX}$ **then**
        $X_{MAX} \leftarrow$ X-axis value of r_c;
    $c \leftarrow c + 1$;
return X_{MAX}

Algorithm 3 FindYMAX(n)

Input: n
Output: The north-most bound of the distribution, Y_{MAX}
$r_i \in R$ considers its position as origin (0,0).
Let Y_{MAX} be Y-value of the north-most robot, and initially, $Y_{MAX} \leftarrow$ Y-value of r_1;
Let c be the robot counter to compare all robots and $c \leftarrow 2$;
while $c <= n$ **do**
  if *Y-axis value of* $r_c > Y_{MAX}$ **then**
    $Y_{MAX} \leftarrow$ Y-axis value of r_c;
  $c \leftarrow c + 1$;
return Y_{MAX}

Algorithm 4 ComputeGrid(n, k)

Input: n, k.
Output: $\{T_1, T_2, \ldots, T_n\}$: Equidistant target points on the grid.
Number of rows and columns of the grid, $rc \leftarrow \lceil\sqrt{n}\rceil$;
X-value of west-most robot,$X_{MIN} \leftarrow$ FindXMIN(n);
X-value of east-most robot,$X_{MAX} \leftarrow$ FindXMAX(n);
Minimum width of the grid, $grid_{width_{min}} \leftarrow X_{MAX} - X_{MIN}$;
Computed minimum distance between adjacent nodes, $x_{length} \leftarrow \lfloor grid_{width_{min}} \div rc \rfloor$;
Given minimum distance to be maintained between two adjacent grid nodes, k;
Thus, the final distance between two adjacent nodes on the final computed grid, $dist_{adj} \leftarrow maximum(x_{length}, k)$;
Width of the final computed grid, $grid_{width} \leftarrow dist_{adj} * rc$;
Y-axis value of the north-most robot, $Y_{MAX} \leftarrow$ FindYMAX(n)
Let, the north-west corner target position of the grid be T_1 and
$T_1 \leftarrow$ coordinates (X_{MIN}, Y_{MAX});
$row = 1$;
$column = 1$;
$i = 2$;
while $row \leq rc$ **do**
  while $column \leq rc$ **do**
    T_i = A point on the grid, $dist_{adj}$ unit apart from T_{i-1};
    $i = i + 1$;
    $column = column + 1$;
  $row = row + 1$;
To ensure the agreement in grid computation of the robots, the boundary robots move to their nearest vacant target positions on their respective boundary and get fixed;
If there is any other overlapping robot(s) on this nearest target position, the other robot(s) slides(s) out of this target position. Thus,
Extreme north robot(s) on Y_{MAX} bound move to their nearest vacant target on Y_{MAX};
Extreme west robot(s) on X_{MIN} bound move to their nearest vacant target on X_{MIN};
Extreme east robot(s) on Y_{MAX} bound move to their nearest vacant target on X_{MAX}.
return $\{T_1, T_2, \ldots, T_n\}$;

4.4 Description of the Algorithm FormGrid

The robots compute and move to the target nodes of the grid using FormGrid. They start from the north-west corner of the grid.

At first, to ensure agreement in grid computation of the robots, the boundary robots move to their nearest vacant target positions on their respective boundary and get fixed. If there is any other overlapping robot(s) on this nearest target position, the other robot(s) slides(s) out of this boundary target position. Therefore, the 'ComputeGrid' routine always agrees on the same $grid_{width}$ and thus the same $T_1, T_2, ..., T_n$.

Next if there is any robot overlapping a target point, it slides to that target point. In case are multiple overlapping, the north-most among them slides to the target point. If there are two robots with the same Y-axis value, overlapping a target position, then the west one with the less X-axis value than the other, moves to the target point while the other slides out of that target position.

The unoccupied grid positions are then filled one by one starting from the north-west corner of the grid, i.e. the target position with the maximum Y-axis value and minimum X-axis value T_i is filled first. If it is already occupied, then the next target position on the grid with the next lowest X-axis value on the same row (i.e. same Y-axis value) T_{i+1} is considered.

The robot r_i nearest to the centre of the target point T_i moves to this destination point T_i. If there is more than one robot closest to this target point T_i, then the one with the maximum Y-axis value is considered to move to the destination T_i. If there are two nearest robots with the same Y-axis value, the west one with the lesser X-axis value is considered to move to the target point T_i. This robot r_i is fixed at this point and does not move further.

The extreme west robot of a row row_x on the grid is marked as $left_{rx}$. The unoccupied target positions from row_x are considered one by one, and the robot nearest to the target position moves to it. Once the robots encounter that the extreme east point on the grid ($grid_{width}$ from $left_{rx}$) is reached, the whole row_x is considered filled. After the row_x is completely filled, the unplaced robots (i.e. robots that are not in their target points on the grid) between row_{x-1} and row_x are identified (say $r_{i_{x-1,x}}$). Now, these robots are the ones that try to get placed. The next row row_{x+1} then starts to be filled, from the west side, such that the robot $r_{i_{x-1,x}}$ nearest to a target node T_i moves to that target node, and in case of two such robots, the west one of them with less X-axis value moves to the target position.

The possibilities that can occur here and the way they are handled are as follows:

- Case 1: No unoccupied target point T_i in row_x but robot $r_{i_{x-1,x}}$ between row_{x-1} and row_x still left to be placed in a target position.
  Here, the robot $r_{i_{x-1,x}}$ between row_{x-1} and row_x moves southwards along its Y-axis to the intersection point at row_x. If $r_{i_{x-1,x}}$ finds another robot $r_{j_{x-1,x}}$ in its path vertically downwards to row_x, then it stops and wait for $r_{j_{x-1,x}}$ to place itself on row_x. Then $r_{i_{x-1,x}}$ moves to row_x at the next available unoccupied location on row_x. If there is no available location on row_x, then $r_{i_{x-1,x}}$ waits till there is space

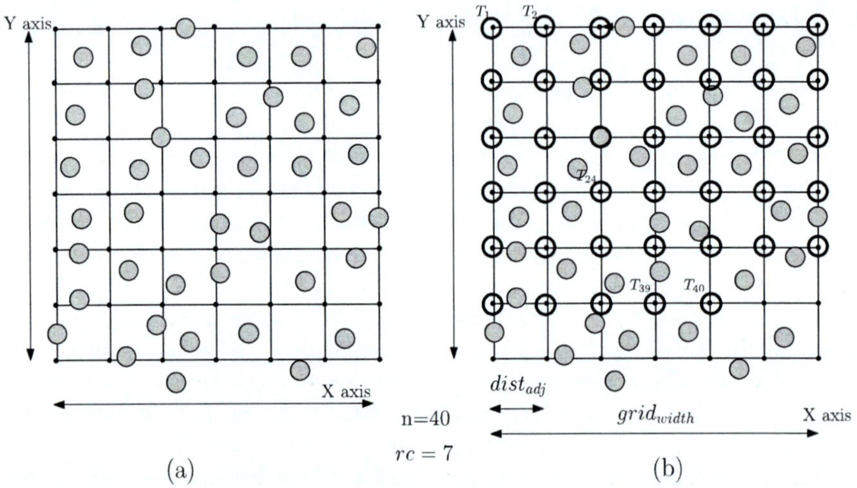

Fig. 2 An example of a grid

available on the row_x after any of the already occupying unplaced robots on row_x is placed at its target point on the further rows.

- Case 2: No $r_{i_{x-1,x}}$ left between row_{x-1} and row_x, but still there are unoccupied target points T_i in row_x.
  In this case, the next unoccupied target point T_i is considered, and the robot $r_i \in$ R nearest to it moves to T_i following the same rule, i.e. if there are more than one such $r_i \in$ R then the one north-most amount them is considered. Furthermore, if there are two such $r_i \in$ R nearest to T_i and having the same Y-axis value, the one on west with lesser X-axis value is considered to move to T_i. Thus, in a similar way, the target points are filled one by one from north to south and from west to east.
- Case 3: Neither any robot $r_{i_{x-1,x}}$ left unplaced between row_{x-1} and row_x nor any unoccupied target point T_i left in row_x.
  In this case, no further action is needed on row_{x-1} and row_x as they are already filled, and there are no $r_{i_{x-1,x}}$ between row_{x-1} and row_x. So, the next row, i.e. row_{x+1} is considered now and the unoccupied target points on row_{x+1} starting from the west-most one starts filling in the same way.

Thus, every time a row is completely filled with no vacant target points, it is checked if there is any unplaced robot $r_{i_{x-1,x}}$ between row_{x-1} and row_x. These robots try to move southwards to row_x. Also while occupying the target points of row_x, highest priority is given to the robots that are north of row_x. This process continues until all the robots reach their target points on the grid (Fig. 3).

Finally, the grid is uniformly filled with the top rows completely filled, and the last row may be completely or partially filled. The grid with rc number of rows and columns may or may not be completely filled depending on the number of available

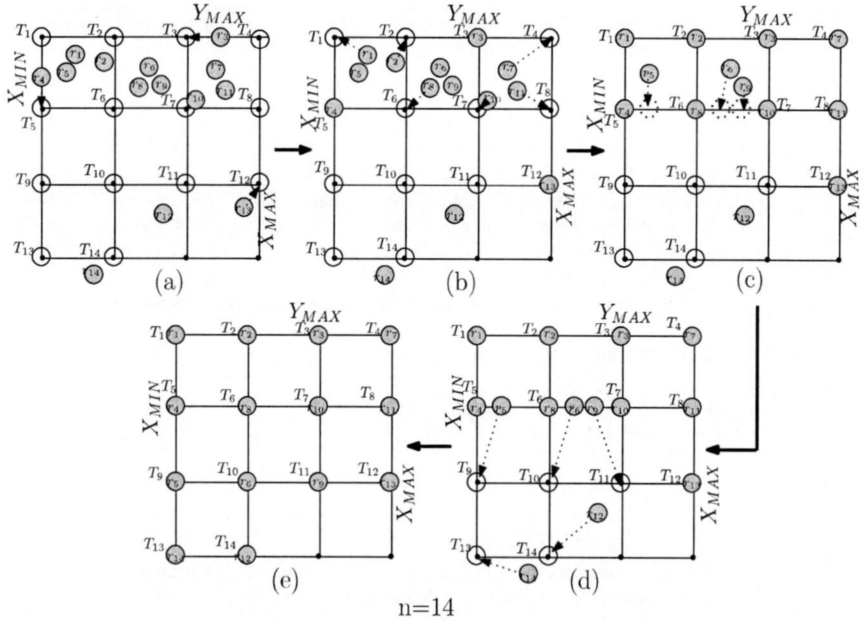

Fig. 3 Grid with n = 14 robots

robots, n. Algorithm *FormGrid* forms the grid by transferring the robots from their initial positions to their target positions on the grid (Algorithm 5).

Definition 2 A vacant target position is a point on the grid, such that a circular zone of radius unit, around this point, contains no part of another robot.

Algorithm 5 FormGrid(n, k)

Input: n,k
Output: A robot $r_i \in R$ moves to its target node on the grid
Let T be a target position on the grid (i.e. grid node)
Initially, r_i is at its local coordinate system origin, i.e. (0.0)
$\{T_1, T_2, \ldots, T_n\} \leftarrow ComputeGrid$(n,k);
if $\exists$ r_i *overlapping* T_i **then**
  if $\exists$ *single* r_i *overlapping* T_i **then**
    r_i slides to T_i;
  if $\exists$ *multiple robots overlapping* T_i **then**
    if $\exists$ *a single overlapping* r_i *with maximum Y-axis value* **then**
      r_i with maximum Y-axis value sides to T_i;
    if $\exists$ *two overlapping robots with same maximum Y-axis value* **then**
      r_i with the lesser X-axis value moves to T_i while the other slides out of T_i;
T_i ← vacant target point with the maximum Y-axis value and minimum X-axis value.
The robot r_i nearest to T_i moves to destination T_i;
x=1; y=1;
i=1;

```
while x ≤ rc do
    while y ≤ rc do
        if T_i is vacant then
            if (x − 1) > 0 and ∃ r_i between row_x and row_{x−1} then
                if ∃ single robot r_i nearest to T_i and between row_x and row_{x−1} then
                    r_i moves to T_i;
                if ∃ multiple robots {r_1, ..., r_k} nearest to T_i and between row_x and row_{x−1} then
                    if ∃ a single nearest robot r_i between row_x and row_{x−1} with maximum Y-axis
                    value nearest to T_i then
                        r_i with maximum Y-axis value moves to T_i;
                    if ∃ two nearest robots with same maximum Y-axis value between row_x and
                    row_{x−1} then
                        r_i with the lesser X-axis value moves to T_i;
            if x = 1 or ∃ no r_i between row_x and row_{x−1} then
                if ∃ single robot r_i nearest to T_i then
                    r_i moves to T_i;
                if ∃ multiple robots nearest to T_i then
                    if ∃ a single nearest robot r_i with maximum Y-axis value nearest to T_i then
                        r_i with maximum Y-axis value moves to T_i;
                    if ∃ two nearest robots with the same maximum Y-axis value then
                        r_i with the lesser X-axis value moves to T_i;
        i=i+1; y=y+1;
    if (x − 1) > 0 and ∃ r_i between row_x and row_{x−1} then
        if r_i finds no other robot r_j in its path along T axis to row_x then
            if ∃ vacant space available at the intersection of r_i's Y-axis and row_x then
                r_i moves southwards along its Y-axis to row_x and waits;
            r_i moves to row_x at the next available unoccupied location on row_x;
        r_i stops and wait for r_j to place itself on row_x;
    x=x+1;
return {T_1, T_2, ..., T_n};
```

Lemma 1 *When a autonomous robot r_i is moving towards T_i, no other robot is obstructing its path, i.e. r_i's movement is collision-free.*

Proof In this grid formation algorithm, based on the ordering and priority listed, the nearest robot to an empty destination node moves to that node. For occurrence of any obstruction by other robots, following two situations may occur:

- The obstacle robot is closer to the vacant target position. This is impossible as then the obstacle robot will be considered to move to the target point.
- When row$_x$ is completely filled, if there still exists any robot $r_{i_{x,x-1}}$ between row$_x$ and row$_{x-1}$, these robots move down vertically to row$_x$. In this case, it may so happen that there is another robot $r_{j_{x,x-1}}$ along this path down to row_x. Here, $r_{i_{x,x-1}}$ does not move and wait till $r_{j_{x,x-1}}$ has moved to row$_x$. $r_{i_{x,x-1}}$ then moves to row_x at a next available vacant location on row$_x$.

Therefore, the robots arrive at their respective destinations without encounter in any collision.

Lemma 2 *The algorithm is free of deadlock. Hence, progress is assured.*

Proof the fat mobile robots start occupying the empty node positions, from the north-west corner of the computed grid. As the target positions or nodes of the grid is always greater than or equal to the number of robots in R, there invariably exists an empty target node to be occupied by its nearest robot based on the priorities listed. The ordering of the movement of the robots is implicitly managed by this algorithm to avoid any tie. Hence, there is no deadlock.

Thus, we derive the below theorem.

Theorem 1 *A group of autonomous, homogeneous, oblivious, asynchronous, transparent fat robots can deterministically form an uniform grid in finite time, under unlimited visibility, agreement on orientation and direction of the axes.*

Proof The robots in this model are autonomous mobile robots as they can operate independently by performing cycles of the phases 'wait–look–compute–move'. These are homogeneous mobile agents and cannot be distinguished from each other. Thus, the robots take photograph in the 'look' phase to identify the positions of the other robots at that point of time. In this algorithm, the robots do not need to recollect any past action or looked data from an earlier cycle, i.e. they are oblivious. Here the robots have asynchronous scheduler where the robots perform their tasks independently. Since at a given time, the robot nearest to a given node is activated, and also there is ordering given in case of multiple such robots, as per the algorithm *FormGrid*, a single robot will move be activated and move. Thus, an asynchronous scheduler will work fine. The transparent fat robots have full visibility, as they need to determine the dimensions and bounds of the grid distribution to be formed. The north bound (i.e. Y_{MAX} robot, west bound (i.e. X_{MIN} robot) and east bound (i.e. X_{MAX} robot) of the grid to be formed are fixed in *ComputeGrid* algorithm, and thus, the determination of the grid dimensions is consistent for all the robots at all times. Finally, we have assumed that though the robots have local coordinate system, they have common perception on the direction of the axes. This assumption ensures that all robots on the plane, compute and identify the same robot(s) as the north-bound Y_{MAX} robot(s) at any point of time. Similarly, they identify the same robots as the west-bound X_{MIN} robot and the east-bound X_{MAX} robot, respectively. With this underlying model, from lemma 2 we have a deadlock-free algorithm that assures progress and from lemma 3, we guarantee no collision of robots, thus deterministically reaching the goal of the algorithm, i.e. fat robots can deterministically form a uniform grid in finite time.

5 Conclusion

In this paper, the formation of grid problem with oblivious transparent fat robots has been addressed. Collision avoidance is the primary objective of the algorithm, and it has been addressed successfully. The paper presents a distributed algorithm that assumes asynchrony, rigidity of movement of the robots and axes agreement on direction and orientation. Our proposed algorithm converges in finite time. However, we have used transparent fat robots to achieve full visibility. The analysis of complexity for the proposed algorithm remains a future work. The immediate extension of this work would be to consider opacity of the robots. Investigation of the problem with limited visibility is another direction.

References

1. Ackerman E, Robots that can efficiently disinfect hospitals using uv light could slow coronavirus infections. https://spectrum.ieee.org/automaton/robotics/medical-robots/autonomous-robots-are-helping-kill-coronavirus-in-hospitals. Accessed: 2020-08-09
2. Barrameda EM, Das S, Santoro N (2008) Deployment of asynchronous robotic sensors in unknown orthogonal environments. In: Fekete SP (ed) Algorithmic aspects of wireless sensor networks. Berlin, Heidelberg. Springer Berlin Heidelberg, pages 125–140
3. Barriere L, Flocchini P, Mesa-Barrameda E, Santoro N (2009) Uniform scattering of autonomous mobile robots in a grid. In *2009 IEEE International Symposium on Parallel Distributed Processing*, pages 1–8, May 2009
4. Bose K, Adhikary R, Kundu MK, Sau B (2020) Arbitrary pattern formation on infinite grid by asynchronous oblivious robots. Theoret Comput Sci 815:213–227
5. Cao YU, Fukunaga AS, Kahng AB, Meng F (1995) Cooperative mobile robotics: antecedents and directions. In: Proceedings 1995 IEEE/RSJ international conference on intelligent robots and systems. Human robot interaction and cooperative robots, volume 1, pages 226–234
6. Chaudhuri SG, Mukhopadhyaya K (2016) Distributed algorithms for swarm robots. Control, and modeling of swarm robotics. In: Handbook of research on design
7. Cohen R, Peleg D (2008) Local spreading algorithms for autonomous robot systems. Theoret Comput Sci 399(1):71–82. Structural Information and Communication Complexity (SIROCCO 2006)
8. Czyzowicz J, Gasieniec L, Pelc A (2006) Gathering few fat mobile robots in the plane. In: Shvartsman MMAA (ed) Principles of distributed systems. Berlin, Heidelberg, 2006. Springer, Berlin Heidelberg, pages 350–364
9. DST. Uv disinfection trolley can effectively clean up hospital spaces to combat covid-19. https://vigyanprasar.gov.in/vigyan-samachar/. Accessed: 2020-08-09
10. Défago X, Souissi S (2008) Non-uniform circle formation algorithm for oblivious mobile robots with convergence toward uniformity. Theoret Comput Sci 396(1):97–112
11. Efrima A, Peleg D (2007) Distributed algorithms for partitioning a swarm of autonomous mobile robots. In: Prencipe G, Zaks S (eds) Structural information and communication complexity. Berlin, Heidelberg. Springer Berlin Heidelberg, pages 180–194
12. Flocchini P, Prencipe G, Santoro N, Widmayer P (2008) Arbitrary pattern formation by asynchronous, anonymous, oblivious robots. Theoret Comput Sci 407(1):412–447
13. Heo N, Varshney PK (2003) A distributed self spreading algorithm for mobile wireless sensor networks. In: 2003 IEEE wireless communications and networking, 2003. WCNC 2003, vol 3, pages 1597–1602
14. Hsiang T-R, Arkin EM, Bender MA, Fekete SP, Mitchell JSB (2004) Algorithms for rapidly dispersing robot swarms in unknown environments. Springer, Berlin Heidelberg, Berlin, Heidelberg, pp 77–93
15. Kasuya M, Ito N, Inuzuka N, Wada K (2006) A pattern formation algorithm for a set of autonomous distributed robots with agreement on orientation along one axis. Syst Comput Jpn 37(10):89–100

16. Katreniak B (2005) Biangular circle formation by asynchronous mobile robots. In: Pelc A, Raynal M (eds) Structural information and communication complexity. Berlin, Heidelberg. Springer Berlin Heidelberg, pages 185–199
17. Kshemkalyani AD, Molla AR, Sharma G (2020) Dispersion of mobile robots on grids. In: Rahman MS, Sadakane K, Sung W-K (eds) WALCOM: algorithms and computation. Springer International Publishing, Cham, pp 183–197
18. Li X, Santoro N (2006) An integrated self-deployment and coverage maintenance scheme for mobile sensor networks. In: Cao J, Stojmenovic I, Jia X, Das SK (eds) Mobile Ad-hoc and sensor networks. Berlin, Heidelberg. Springer Berlin Heidelberg, pages 847–860
19. Payton D, Estkowski R, Howard M (2005) Pheromone robotics and the logic of virtual pheromones. In: Şahin E, Spears WM (eds) Swarm robotics. Berlin, Heidelberg. Springer Berlin Heidelberg, pages 45–57
20. Poudel P, Sharma G (2019) Time-optimal uniform scattering in a grid. In: Proceedings of the 20th international conference on distributed computing and networking, ICDCN '19, New York. Association for computing machinery, page 228–237
21. Poudel P, Sharma G (2020) Fast uniform scattering on a grid for asynchronous oblivious robots. In: Devismes S, Mittal N (eds) Stabilization. Safety, and security of distributed systems. Springer International Publishing, Cham, pp 211–228
22. Schranz M, Umlauft M, Sende M, Elmenreich W (2020) Swarm robotic behaviors and current applications. Front Robot AI 7:36
23. Simon M, The covid-19 pandemic is a crisis that robots were built for. https://www.wired.com/story/covid-19-pandemic-robots/. Accessed: 2020-08-09
24. Yamauchi Y, Yamashita M (2013) Pattern formation by mobile robots with limited visibility. In: Moscibroda T, Rescigno AA (eds) Structural information and communication complexity. Springer International Publishing, Cham, pp 201–212

Chapter 8
A LSB Substitution-Based Steganography Technique Using DNA Computing for Colour Images

Subhadip Mukherjee, Sunita Sarkar, and Somnath Mukhopadhyay

Abstract Steganography is the process of using a cover or medium such as photograph, audio, text and video to shield information from the outer world. This paper suggested a new approach based on DNA computing for hiding information within an image using the least significant bit (LSB). To do this, the DNA is decomposed by four nucleotides, namely adenine, thymine, guanine and cytosine. Whereas a codon is a sequence of three nucleotides and to represent a codon, the two-pixel LSBs from the image are taken and then converted into protein. The confidential data bits are concealed into the codons, which transmutes the original cover image into a stego-image which is completely trustworthy to avoid human visual system, and the confidential data is impossible to detect. The empirical findings show the effectiveness of the suggested approach by producing 0.750 bpp of hiding power with average 56.48 dB of peak signal-to-noise ratio (PSNR) which makes it a strong image steganography technique.

Keywords Image steganography · Data hiding · DNA computing · LSB substitution

1 Introduction

The need for encrypted contact is growing rapidly day by day and has existed since the humans started to communicate electronically. At this Internet era, the sharing of information has led to a rapid rise in the transformation of information between the sender and the recipient. Although transmitting information through the Internet has made life easier, it struggles from a critical challenge, i.e. secure data transfer [1, 2]. This is all attributed to the growing amount of cases of hacking and interference last year. Two key techniques, including cryptography [3, 4] and steganography

S. Mukherjee (✉)
Department of Computer Science, Kharagpur College, Kharagpur 721305, India

S. Sarkar · S. Mukhopadhyay
Department of Computer Science and Engineering, Assam University, Silchar, India

J. K. Mandal et al. (eds.), *Proceedings of International Conference on Innovations in Software Architecture and Computational Systems*, Studies in Autonomic, Data-driven and Industrial Computing, https://doi.org/10.1007/978-981-16-4301-9_9

[5, 6], have come into being and have been used widely around the world to solve reliable communication problems. However, to build more robust processes, they can be mixed and this makes it more difficult to crack the protection [7]. However, cryptography requires ways of encrypting a private information to render it strongly and it is suspicious. However, steganography is a process that masks a private information within an original multimedia file without perverting any information about the file used to make it less suspicious. It is possible to use different media forms, such as image, audio, video and DNA sequences [8, 9]. The distortion of the cover media to an extended degree may give any steganalysis attacker a clue that such media contains a confidential information and therefore the attackers may attack or destroy the cover media to detect or destroy the confidential data. In order to hide a data, different media such as video, audio and image can be applied to construct a more secure and robust technique. Since image is the most popular multimedia used through the Internet that is why we have used image to build our proposed technique and to achieve a high-quality stego-image, the DNA computing-based techniques are remarkable. Therefore, in this paper, a secure and robust image steganography technique based on DNA computing is suggested to generate a stego-image with high PSNR value.

The remaining portions of this paper are structured as follows: In Sect. 2, the works related to the suggested technique are discussed and after that the suggested methodology containing both the embedding and extraction procedures is discussed in Sect. 3. The analysis of the outcomes of the suggested methodology is discussed in Sect. 4. Lastly, in Sect. 5, the conclusion is depicted.

2 Related Works

Over the years, several strategies have been created to conceal private data within an image file and out of those techniques, the LSB-based techniques are very commonly, easily and popularly used techniques [10, 11]. The DNA computing technique has attracted many scholars, researchers and other academicians in the research field of information hiding [12, 13]. The main three reasons behind attracting the attention of the researchers in DNA steganography are: (1) invisibility, hiding steganalysis results, (2) hiding capability, having a fair size of data embedding capability and (3) robustness [14]. The arrangement of nucleotides is either modified or adjusted in order to conceal a message in the DNA chain. The nucleotides are translated to an equivalent stream of bits from the English letters for the simplicity of the process, and this concept was introduced and implemented in the year 2000 by Leier [15]. Based on some mathematical models, some researchers [16, 17] have done several works on the randomization of the nucleotides (A, G, T, C). Moreover, too much modifications of the DNA nucleotides could make codons suspicious to the attackers. However, if the RNA translation takes place to generate the proteins, the codons can be damaged its functionality and hence the extraction will not work.

Based on two widely used data hiding methods, lossless compression and difference expansion (DE), Chen [18] developed a technique. According to this scheme, they coded the private data into the DNA nucleotides (ACTG) form using 2-bit binary representation. After that, the equivalent sequence of decimal digits formed by grouping k bits was generated and is classified into two sets of pairs of decimal digits; one set is known as expandable set ($C1$), and the other set is known as changeable set ($C2$). Based on these sets, the compressed location map is to be constructed to conceal the private data within the original LSBs of pairs using difference expansion. Though they have achieved a data hiding capacity of 0.13 bpn, it is excessively low for these days. The steganography technique suggested by Liu et al. [19] is established with the help of linear chaotic map which is also piecewise. Image steganography method for tamper restoration with the help of DNA sequences was explained by Fu et al. [20]. Though, these schemes are neither capable nor suitable for producing a high-quality stego-image. Therefore, the proposed technique is developed to overcome the above problems.

3 The Proposed Scheme

The proposed concept of image steganography is mainly based on two data embedding techniques; one is the LSB technique, and the other is DNA sequencing to make a high-quality stego-image. In Fig. 2, the mechanism of the proposed steganography scheme is depicted. The corresponding 2D matrices for blue, green and red are retrieved from the original image. In order to achieve the objective of high PSNR value, the DNA is decomposed of protein synthesis (see Fig. 1) and the properties of this decomposition are translated and serialized. Due to the binary representation of the cover image LSBs for RGB, therefore, a method must be formed to map these bits with the nucleotides. The suggested strategy would then use the binary representation of nucleotides C, T, G, A (see Table 1). With reference to Table 2, the data embedding procedure injects the private bits into the LSBs of a selected colour 2D matrix and produces corresponding stego message. In Sect. 3.2, the complete extraction mechanism is explained to obtain the confidential information and

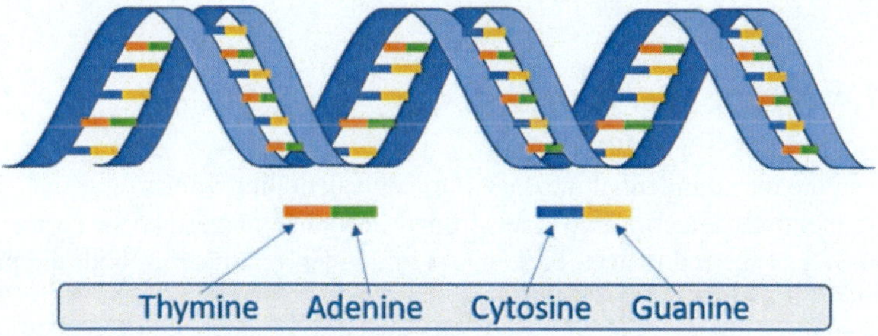

Fig. 1 Basic structure of a DNA

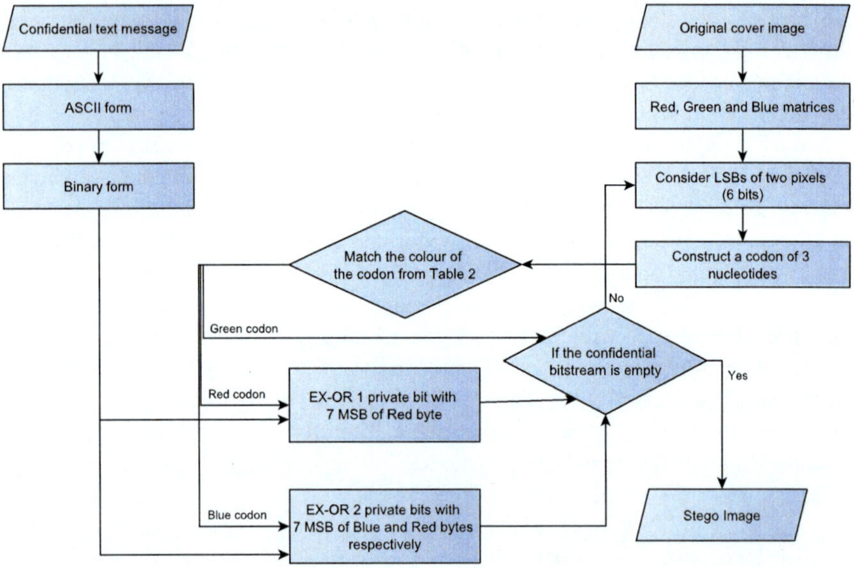

Fig. 2 Entire scenario of the proposed image steganography methodology

Table 1 Nucleotides with corresponding codes

Nucleotides	Short form	Corresponding code
Cytosine	C	11
Thymine	T	10
Guanine	G	01
Adenine	A	00

the interesting fact is that the extraction mechanism follows just the opposite steps followed by the embedding procedure which is depicted in Sect. 3.1. Moreover, these procedures make this suggested technique more harder to track out the existence of the confidential data hidden within the cover image (Fig. 2).

3.1 *Embedding Procedure*

1. Transmute the confidential message into equivalent binary form.
2. By sequentially selecting two pixels from the original image, take the corresponding LSBs of the red matrix, blue matrix and green matrix from both the pixels, i.e. 3 + 3 = 6 bits.
3. Split these 6 bits in 2 + 2 + 2 form and encapsulate these 6 bits to compose a codon with reference to Table 1.

Table 2 Colour grouping of codons

Colour of codons	Codons
Set of blue codons	ACT, ACC, ACG, ACA, CCT, CCC, CCG, CCA, GCT, GCC, GCG, GCA, GGT, GGC, GGG, GGA, GTT, GTC, GTG, GTA, TTA, TTG, ATC, ATT, ATA, CGT, CGA, CGC, CGG, TCT, TCC, TCG, TCA, CTT, CTC, CTG, CTA
Set of red codons	AAA, AAG, AAC, AAT, TGC, TGT, CAA, CAG, GAC, GAT, CAC, CAT, TAC, TAT, GAA, GAG, TTC, TTT, AGA, AGG
Set of green codons	AGA, AGG, TAA, TAG, TGA

4. Now, by mapping the colours of the codons with reference to Table 2, for blue colour, conceal 2 bits by applying the EX-OR operation with seven MSBs of blue and red matrices of second pixel, respectively. For the matrix of red, embed 1 bit by applying the EX-OR operation between the seven MSBs of the red matrix of second pixel. But, for the green colour matrix, jump to the next step.
5. Repeat the previous three steps until the entire bit stream is concealed.

3.2 *Extraction Procedure*

1. By sequentially selecting two pixels from the stego-image, take the corresponding LSBs of the red matrix, blue matrix and green matrix from both the pixels, i.e. 3 + 3 = 6 bits.
2. Split these 6 bits in 2 + 2 + 2 form and encapsulate these 6 bits to compose a codon with reference to Table 1.
3. Now, for blue colour, extract 2 bits by applying the EX-OR operation with seven MSBs of blue and red matrices of second pixel, respectively. For the matrix of red, extract 1 bit by applying the EX-OR operation between the seven MSBs of the red matrix of second pixel. But, for the green colour matrix, jump to the next step.
4. Repeat the previous three steps until all the private bits are extracted.
5. Combine all the extracted bits and transmute into the required message.

4 Experimental Analyses

The suggested method is experimented with 4 test images, namely tree, baboon, airplane and Lena from the USCID database [21] in MATLAB R2011a. The images taken for the experiments are of size 256 × 256 and are shown in Fig. 3. Here, the metrics, normalized absolute error (NAE) and peak signal-to-noise ratio (PSNR), are

Fig. 3 Test images

evaluated to judge security as well as the visual quality of stego-image. The most common picture quality parameter used for stego picture quality evaluation is the PSNR. The PSNR for the original image $V_{c,d}$ of size $C \times D$ and stego-image $V'_{c,d}$ is defined in Eq. (1).

$$\text{PSNR} = 20\log_{10}\frac{255}{\sqrt{\frac{1}{CD}\sum_{c=1}^{C}\sum_{d=1}^{D}(V_{c,d} - V'_{c,d})^2}} \tag{1}$$

Normalized absolute error is an important metric for determining the robustness of the proposed scheme, and it is defined in Eq. (2).

$$\text{NAE} = \frac{\sum_{c=1}^{C}\sum_{d=1}^{D}(V_{c,d} - V'_{c,d})}{\sum_{c=1}^{C}\sum_{d=1}^{D}V_{c,d}} \tag{2}$$

The embedding capacity of the suggested technique $= \frac{\text{total number of concealed bits}}{\text{number of pixels of the image}} = \frac{49,152}{256 \times 256} = \frac{49,152}{65,536} = 0.750$ bits per pixel.

The outcomes with respect to the metric PSNR, NAE and embedding capacity (EC) of the suggested technique are displayed in Table 3. We all know that a scheme is said to be a good steganography scheme if it can achieve more than 30 dB of PSNR. From this table, it is clear that we have achieved a remarkable PSNR value of 56.48 dB on average, which is 88.27 % higher than the standard value, with the EC of 0.750 bpp. The value of NAE closer to zero means the error is closer to zero and the human visual system cannot detect the stego-image. The proposed scheme has achieved average 0.0018 of NAE value. Therefore, we can say that the proposed technique not is only capable of producing higher PSNR but also can provide higher visual security.

To establish the potentiality of the suggested scheme, we have compared our scheme with other popular and recent DNA computing-based image steganography schemes. In Table 4, average EC and average PSNR of Muhammad [22], Nag [23] and Zhang [24] are compared with the proposed methodology. From this table, it must be espy that the proposed technique has achieved the average PSNR of 56.48

Table 3 Outcomes of the suggested technique

Image	PSNR (dB)	NAE	Embedding capacity (bpp)
Tree	56.49	0.0017	0.750
Airplane	56.47	0.0019	0.750
Lena	56.49	0.0016	0.750
Baboon	56.46	0.0018	0.750

Table 4 Result comparison with other schemes

Scheme	Metric	Value
Muhammad et al.[22]	PSNR (dB)	53.89
	Embedding capacity (bpp)	0.500
Nag et al.[23]	PSNR (dB)	55.44
	Embedding capacity (bpp)	0.700
Zhang et al.[24]	PSNR (dB)	52.92
	Embedding capacity (bpp)	0.0.300
Proposed	PSNR (dB)	56.48
	Embedding capacity (bpp)	0.750

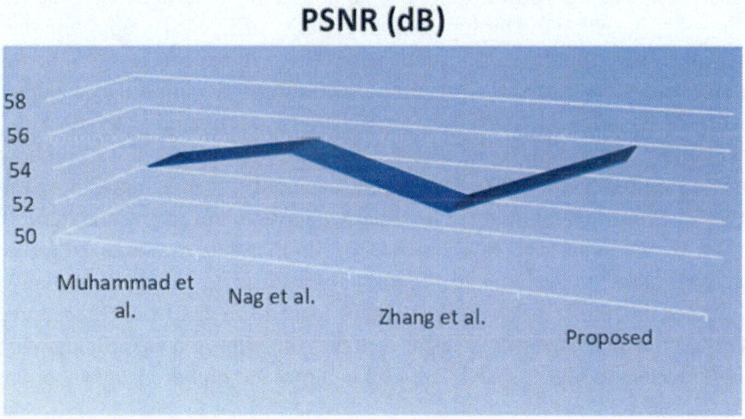

Fig. 4 PSNR comparison

dB which is 2.59, 1.04 and 3.56 dB higher than Muhammad [22], Nag [23] and Zhang [24] (see Fig. 4), respectively. The proposed technique has also achieved the embedding capacity 0.750 bpp which is 0.250, 0.050 and 0.450 bpp higher than Muhammad [22], Nag [23] and Zhang [24] (see Fig. 5), respectively.

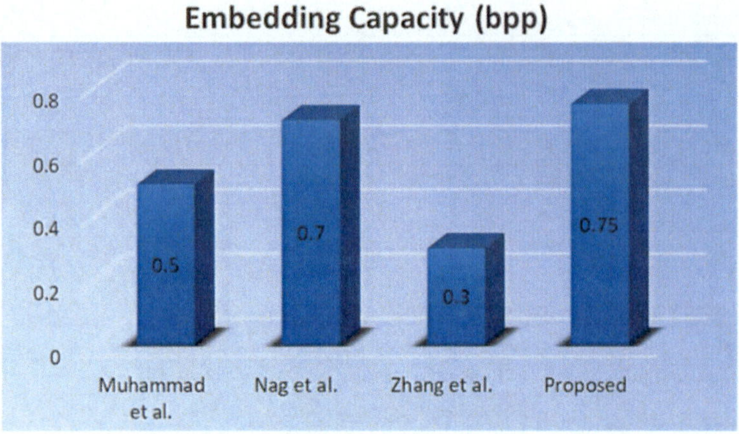

Fig. 5 EC comparison

5 Conclusion

A new image steganography technique based on the deoxyribonucleic acid computing is proposed in this paper. Unlike conventional image steganography approaches, this methodology is based on the biological functionality of DNA. By applying the LSB technique, the secret bits are concealed within the original image with the help of the codons converted into series of amino acids from the DNA. To establish the potentiality of the suggested scheme, our scheme is compared with other popular and recent DNA computing-based image steganography schemes. We all know that a scheme is said to be a good steganography scheme if it can achieve more than 30 dB of PSNR. From this table, it is clear that we have achieved a remarkable PSNR value of 56.48 dB on average, which is 88.27 % higher than the standard value. The value of NAE closer to zero means the error is closer to zero and the human visual system cannot detect the stego-image. The proposed scheme has achieved average 0.0018 of NAE value. Therefore, it is proved that the proposed technique not only is capable of producing higher PSNR but also can provide higher visual security.

References

1. Rani SS, Alzubi JA, Lakshmanaprabu S, Gupta D, Manikandan R (2019) Optimal users based secure data transmission on the internet of healthcare things (Ioht) with lightweight block ciphers. Multimedia Tools and Applications, pp 1–20
2. Gochhayat SP, Lal C, Sharma L, Sharma D, Gupta D, Saucedo JAM, Kose U (2019) Reliable and secure data transfer in Iot networks. Wireless Networks, pp 1–14
3. Easttom W (2021) Quantum computing and cryptography. In: Modern cryptography. Springer, pp 385–390

4. Sadhukhan D, Ray S, Biswas G, Khan M, Dasgupta M (2021) A lightweight remote user authentication scheme for iot communication using elliptic curve cryptography. J Supercomput 77(2):1114–1151
5. Kaur S, Bansal S, Bansal RK (2021) Image steganography for securing secret data using hybrid hiding model. Multimedia Tools Appl 80(5):7749–7769
6. Mukherjee S, Jana B (2019) A novel method for high capacity reversible data hiding scheme using difference expansion. Int J Nat Comput Res (IJNCR) 8(4):13–27
7. Abikoye OC, Ojo UA, Awotunde JB, Ogundokun RO (2020) A safe and secured iris template using steganography and cryptography. Multimedia Tools Appl 79(31):23483–23506
8. El-Khamy SE, Korany NO, Mohamed AG (2020) A new fuzzy-dna image encryption and steganography technique. IEEE Access 8:148935–148951
9. Al-Harbi OA, Alahmadi WE, Aljahdali AO (2020) Security analysis of dna based steganography techniques. SN Appl Sci 2(2):1–10
10. Gambhir G, Mandal JK (2021) Shared memory implementation and performance analysis of lsb steganography based on chaotic tent map. Innovations in systems and software engineering, pp 1–10
11. Chatterjee A, Ghosal SK, Sarkar R (2020) Lsb based steganography with ocr: an intelligent amalgamation. In: Multimedia tools and applications, pp 1–19
12. Nisperos ZA, Gerardo B, Hernandez A (2020) Key generation for zero steganography using dna sequences. In: 2020 12th international conference on electronics, computers and artificial intelligence (ECAI). IEEE, pp 1–6
13. Jose A, Subramaniam K (2020) Dna based sha512-ecc cryptography and cm-csa based steganography for data security. In: Materials today: proceedings
14. Marwan S, Shawish A, Nagaty K (2016) Dna-based cryptographic methods for data hiding in dna media. Biosystems 150:110–118
15. Leier A, Richter C, Banzhaf W, Rauhe H (2000) Cryptography with dna binary strands. Biosystems 57(1):13–22
16. Chang C-C, Lu T-C, Chang Y-F, Lee R (2007) Reversible data hiding schemes for deoxyribonucleic acid (dna) medium. Int J Innov Comput Inf Control 3(5):1145–1160
17. Huang Y-H, Chang C-C, Wu C-Y (2014) A dna-based data hiding technique with low modification rates. Multimedia Tools Appl 70(3):1439–1451
18. Chen T (2007) A novel biology-based reversible data hiding fusion scheme. In: International workshop on frontiers in algorithmics. Springer, pp 84–95
19. Liu G, Liu H, Kadir A (2014) Hiding message into dna sequence through dna coding and chaotic maps. Med Biol Eng Comput 52(9):741–747
20. Fu J, Zhang W, Yu N, Ma G, Tang Q (2014) Fast tamper location of batch dna sequences based on reversible data hiding. In: 2014 7th international conference on biomedical engineering and informatics. IEEE, pp 868–872
21. USCID Image Database. http://sipi.usc.edu/database/
22. Muhammad K, Ahmad J, Farman H, Jan Z (2016) A new image steganographic technique using pattern based bits shuffling and magic lsb for grayscale images. arXiv preprint arXiv:1601.01386
23. Nag A, Choudhary S, Basu S, Dawn S (2016) An image steganography scheme based on lsb++ and rhtf for resisting statistical steganalysis. IEIE Trans Smart Process Comput 5(4):250–255
24. Zhang S, Gao T (2015) A novel data hiding scheme based on dna coding and module-n operation. Int J Multimedia Ubiquitous Eng 10(4):337–344

Chapter 9
An Approach of Safe Stock Prediction Using Genetic Algorithm

Nilanjana Adhikari, Mahamuda Sultana, and Suman Bhattacharya

Abstract Stock market investments are an admired problem but onerous task. Because it is very unstable in nature for different factors and it is very hard to predict the safe stocks to invest at different circumstances. To guess the safe stocks to invest from a very large no of shares is an attractive research area that needs to be done efficiently because it is the question of profit and loss. In this research work, an optimized search algorithm has been used to predict no of the stocks from a huge scale of shares or stocks in which it will be safe to invest. In this proposed work, NIFTY top 50 shares to which it will be safe to invest for long-term investment has been taken into consideration and has been evaluated the safe stocks in a rank wise manner using an efficient search optimization technique, genetic algorithm (GA). Genetic algorithm is capable to yield better accuracy than other similar models. This is a heuristic search optimization method for searching of a very vast domain of dataset. This algorithm reflects the process of natural selection where the fittest individuals are selected for reproduction in order to produce offspring of the next generation. An insignificant population of individual archetypes can successfully search a large space because they comprehend schemes. Beneficial sub-structures that can be theoretically united to make fittest entities. Fitness is determined by investigating a huge number of distinct fitness cases. This procedure can be very effective if the fitness cases also grow by their individual GAs.

Keywords Stock market · Nifty top 50 shares · Genetic algorithm · Fitness function · Safe stock

---

N. Adhikari (✉) · M. Sultana
Techno International New Town, MAKAUT, Kolkata, India

S. Bhattacharya
Guru Nanak Institute of Technology, JIS, Kolkata, India

J. K. Mandal et al. (eds.), *Proceedings of International Conference on Innovations in Software Architecture and Computational Systems*, Studies in Autonomic, Data-driven and Industrial Computing, https://doi.org/10.1007/978-981-16-4301-9_10

1 Introduction

Over the past few years, the stock market has become a vital field of research. An intricate and tricky job in financial markets such as stock market is to foretell the course of direction of financial markets. To search for safe stocks to invest from a number of stocks in stock market is very difficult and many a time it has been watched that not in all the cases the market researchers are predicting the stocks which will be safe to invest for long period of time.

In this composition, an attempt has been made to search from Nifty top 50 shares by an excellent optimization technique, genetic algorithm and has been predicted the safe stocks from Nifty top 50 shares. Here, genetic algorithm has been used rather than the traditional search algorithm because this is an optimization search technique which search on a population rather than from a single point by using more advanced natural selection probabilistic view.

The dataset has been collected and analyzed day wise. The previous close, open price, last price and close price of each stock has been monitored in nifty top 50 shares for 10 years. Each top nifty 50 shares have been assigned with a particular id number from 0 to 49. Each chromosome has been generated by selecting 10 random companies from the group of 50. The initial population containing 40 chromosomes has been generated randomly keeping in mind that no company can appear twice in a single chromosome. Next fitness value has been calculated. Selection of stocks has been done depending upon the fitness value. Crossover has been done on the selected individuals who passed the selection process. There is no predetermined final state (solution) in this work. The terminal condition is end of file or the point where the data will be exhausted. Here mutation has not been done as the population is quite diverse for random selection of the gens. A formal lesson of competitive schematics illustrates that the finest policy for imitating them is to significantly increase their relative fitness. This is proven to be the principle used by genetic algorithms.

Rest of this proposed paper has been ordered as follow. Section 2 describes the related work in this domain. Section 3 describes proposed work as well as theories and algorithms which has been used. In Sect. 4 result has been discussed in detailed. Section 5 delivers conclusion and future work.

2 Related Work

A variety of methods have been built, remodeled and exploited by many associates and specialists. A range of system procedures have been applied on financial time series to get a hold of the trend and predict. In the research work [1] developed a feature transformation method using genetic algorithms which reduces the dimensionality of the feature space and removes irrelevant factors involved in stock price prediction and perform better when compared with linear transformation and fuzzification transformation. Another research has been done on genetic algorithms (GAs) by Kyoung-jae Kim I. H. [2] again to predict stock market to use GA not only to improve the learning algorithm, but also to reduce the complexity of the feature space which reduces dimensionality of the feature space and enhances the generalizability of the classifier. Frank Cross in the research work [3] tries to find relationships that could exist between stock price changes on Mondays and Fridays in the stock market in which it has been observed that prices on Friday have risen more often than any other day and it has also been observed that on Monday the prices have least often risen compared to other days. Altunkaynak in his research work, [4] used genetic algorithm for the prediction of sediment load and discharge. Hyunchul Ahn, K.-j. K [5] suggested that the genetic algorithm can be used to predict in financial bankruptcy which has been also tried to use a similar approach to predict the stock which was used in this experiment is completely novel and looks very promising. In the work [6], it has been taken an attempt to optimize the timing of an automated trader using GA to develop trading rules for short time periods, using Technical Indexes, such as RSI, as GA's chromosome. In the paper [7], the description in this work was too preliminary to allow for a comparison with this proposed system which attempts to generate the buying and selling signals against 30 companies' stocks in Germany (DAX30) in which combination of Technical Indexes applied to GA as well as rank the stocks according to the strength of signals to restructure the portfolio without devise any criterion of profit cashing and loss cutting. In the work [8], it has been taken into consideration that the financial Chinese news has a great influence in stock market in which financial news and technical indicator has been used to optimize the prediction. Next in the work [9], data has been extracted from social media like twitter and Facebook about the sentiments of the stock market and basis on that clustering algorithm has been developed by k-means clustering and genetic algorithm. In the research [10], an integrated stochastic optimization model has been introduced using genetic algorithm. In the paper [11], target price of script has been examined for eight different script and 60 different attributes by the use of Genetic to predict the target price. In the next paper [12], artificial neural network and GA have been used to predict the stock price. In the work [13], a web robot has been used to collect the data from stock market then analyzing those data by the use of genetic algorithm and Support vector machine predicted the stock price. In the research work [14], different features have been selected through the use of genetic algorithm and then the stocks are experimented for predicting stock. In the research work [15] to localize objects, fitness function using genetic programming has been used in which the weighted F-measure of the genetic program has been used with fitness value of the localization of the locations of the identified object. In this work [16], a collaborative system has been designed using genetic algorithm for the prediction of weekly prices movement in the Sao Paulo Stock Exchange Index (Ibo vespa Index); in association a comparative study has also been evaluated with the existing ensemble methods where the performance of this system has been proved to be more accurate. In the next work [17], a Rough Set (RS) system has been employed by using genetic algorithms (GA) to optimize the interval and to predict a particular share "closing price". In the research work [18], genetic algorithm in association with technical indicator has been used to design a method which will help to take decision to "Buy" or "Sell" stocks in Sao Paulo Stock Exchange. In the paper [19], a system has been designed which will prompt for correcting the cut-off value for stop-loss to the Opening Range Breakout (ORB) intraday policy using historical dataset but the parameter set and the solution space was too large, so to optimize that parameter set genetic algorithms has been used. In the next work [20], a combination of fuzzy logic and genetic algorithm has been measured to improve the correctness of the prediction. In the work [21], with the information from twitter of good growth companies, a sentiment has been analyzed using support vector machine (SVM) classifier in which genetic algorithm (GA) has been applied to optimize the trading rule to increase profit. In the research [22], the neural network association with genetic algorithm has been employed to forecast "closing price" of rare earth shares using historical data of the same. In the work [23], GP and RSI have been used to forecast the price of individual share.

Not many have tried to use only genetic algorithms to predict stock prices. Since the genetical algorithm can perform reasonably well in many cases there has to be a way to predict stock price using GA as well. It has been taken, an attempt to search from Nifty top 50 shares and optimized the stocks by the genetic algorithm and has been predicted the safe stocks from Nifty top 50 shares.

3 Proposed Work

NIFTY Next 50 stocks have been computed using free float market capitalization method wherein the level of the index reflects total free float market value of all the stocks in the index relative to a particular base market capitalization value. NIFTY Next 50 Index can be used for a variety of purposes such as benchmarking fund portfolios, launching of index funds, ETFs and structured products. Now, a brief theory about the different term that has been used in this context about genetic algorithm has been discussed as follows.

3.1 Theory

Genetic algorithm is biological processes of reproduction and natural selection to solve for the "fittest" solutions [24]. It is a heuristic search algorithm based on the mechanics of natural selection and genetics and they combine survival of the fittest among string structures to form a search algorithm [25]. In a search optimization technique, when multi-parameter soft and hard constraints is applied then genetic algorithm performs better than the other search technique. The main idea of GA is to start with a population of solutions to a problem, and attempt to produce new generations of solutions which are better than the previous ones [17]. From the fittest parents these genes flow through the generation and sometimes these parents create descendants which is more improved than either parent. In this way after each succeeding generation, they will be more appropriate for their environment.

GA functions through a cycle consisting of four stages: Initial Population, Selection, Crossover, and Mutation. In Initial Population stage, a population of genetic structures, chromosomes are constructed randomly which is encoded by binary strings, real numbers or rules. After this, each chromosome is upgraded by a user-defined fitness function. Then the fittest chromosomes are selected by selection strategies which will be chosen by the user depending upon the nature of the problem. The details of the stages have been described subsequently.

Initial Population

In this composition by the help of NSCPY API, the day wise dataset has been collected from National Stock Exchange from year 2014 to year 2019. Here population size is 50. Chromosome size 10. Each company in the Nifty 50 has been denoted by an index value. Each chromosome is generated by selecting 10 random companies from the group of 50. The initial population containing 40 chromosomes is generated randomly keeping in mind that no company can appear twice in a single chromosome.

Fitness Function

Fitness function has been measured of central tendency calculated keeping in mind opening and closing price of the stock per day change_percentage = 100 * (current_close[i] – previous_close[i]) / previous_close[i] The percent change is the difference between the opening and closing price of a particular company. Every company will have its own value of change percentage.

Selection

Selection has been done based on the result of fitness function. Here, it has been considered any two random chromosomes from the population and has been selected for any one with the higher fitness value using the tournament selection. Thus, the companies having lower fitness will tend to get eliminated.

Crossover

Crossover has been done on the selected individuals who passed the selection process. During crossover, it has been kept in mind that it cannot let the same feature (company) appear in an individual more than once. This might lead to anomalies. So, it has been randomly created a mix of features from all selected individuals and created new individuals and thus making a new population set.

Mutation

This work has been created quite random fashion and random selection for initial population has been done. Here after random selection, it has been randomly created a mixed of features from all selected individuals and created new individuals and thus making a new population set for crossover. So, to introduce mutation in the randomized set it will be more diverse and more distributed result which will be difficult to converge.

Terminal condition

Since there is no predetermined final state (solution) in this problem, the terminal condition is end of file or the point where the data is exhausted. It has been repeated the above process until the terminal condition has met to get the solution.

3.2 Proposed Algorithm

Here, dataset has been collected from NSCPY and saved as a.csv file. It has been collected and analyzed day wise. The different measures like previous close, open price, last price and close price of each stock of nifty top 50 shares has been monitored. Each of these 50 shares has been assigned with id number from 0 to 49 in a general way randomly. Each chromosome has been generated by selecting 10 random companies from the group of 50 shares. The initial population containing 40 chromosomes has been generated randomly keeping in mind that no company can appear twice in a single chromosome. Next fitness value has been calculated. Selection of stocks has been done depending upon the fitness value. Crossover has been done on the selected individuals who passed the selection process. There is no predetermined final state (solution) in this work. The terminal condition is end of file or the point where the data will be exhausted. Here mutation has not been done as the population is quite diverse for random selection of the gens. A formal lesson of competitive schematics illustrates that the finest policy for imitating them is to significantly increase their relative fitness. This is proven to be the principle used by genetic algorithms. The algorithm has been described as follows.

Step 1 Start.

Step 2 Import data using NSCPY.

Step 3 Calculate the daily percentage change in closing price company wise.

Step 4 Building Initial Population.

Step 5 Iterate Step 6 to Step 8 until the data is exhausted.

Step 6 change_percentage = 100 * (current_close[i] – previous_close[i]) / previous_close[i] Where i denote Nifty top 50 shares.

Step 7 Tournament selection executed between the fittest chromosomes.

Step 8 Crossover done keeping in mind that no feature appears in an individual more than once.

Step 9 Stop.

4 Results and Discussions

In the following result, a day wise price change reflection of the Nifty top 50 shares has been reflected. The result shows that after 1000 iteration, the terminal condition arises. Now in the following table toward the left side, it is shown the current fitness

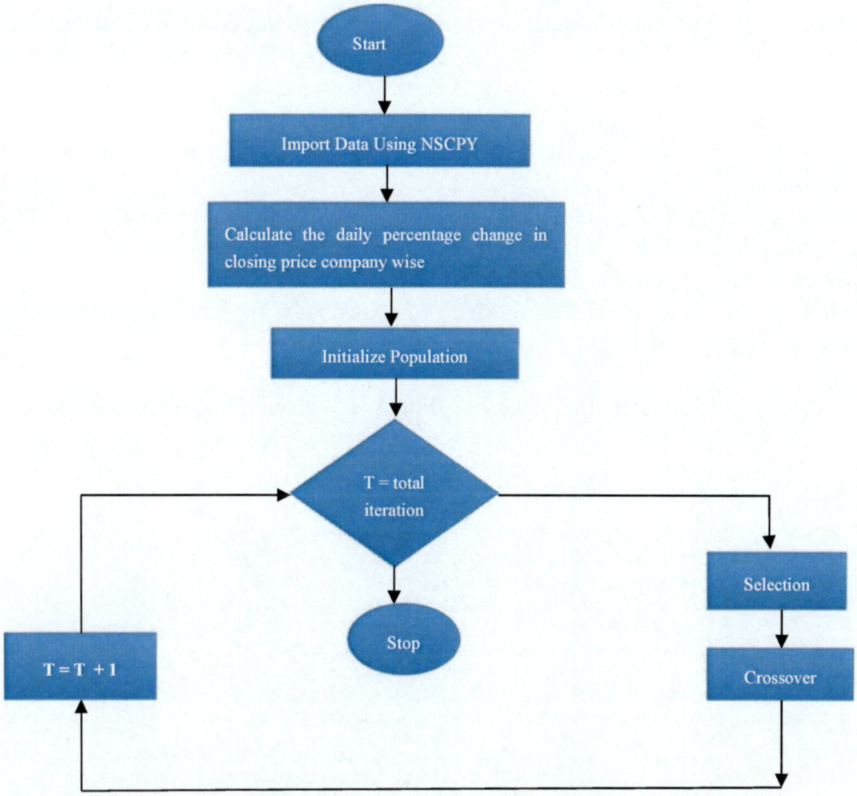

Fig. 1 Synoptic flow of proposed work

value of the particular chromosome in the right side having the company identity no. The graph has been plotted taken into consideration 365 days. Here X-axis reflected as no of days and Y-axis reflected as percentage (%) of change in price. In the graph, it has been noticed that how many times a particular company appears with the highest fitness value. And that company has been taken into consideration as a safe stock to invest from the nifty top 50 shares. A period consisting of 10 years has been taken into account to study the underlying trend in the market. In each iteration there is a progressive improvement of the objective function. Here five iteration or output has been measured, and as per the genetic algorithm every time, it is not showing the exact result everytime but the consistency of the result is more or less same. Here if more and more constraints can be added then the result would be more consistent for all the time.

Iteration 1

Population after 1000 iterations (Fitness of chromosome [chromosome]).

In the following Table 1, in the row-1 company-id 19 has come twice and so on. Now here the highest fitness is 14.0129 and it has been noticed that company-id 6 has come three times and company-id 19 has come twice so as per the Iteration-1 it is safe to invest in company no 6 and 19 respectively.

In Fig. 1, there is a graph stating which company has come how many times.

Iteration 2

Final Population after 1000 iterations (Fitness of chromosome [chromosome]).

In the followingTable 2, in the row-1 company-id 33 has come three and so on. Now here the highest fitness is 15.422.

In Fig. 2, there is a graph stating which company has come how many time bases on the result of Iteration-2.

Iteration 3

Final Population after 1000 iterations (Fitness of chromosome [chromosome]).

Table 1 Frequency of occurrence of each company in iteration-1

```
(8.33796920841103, [2, 5, 19, 19, 22, 26, 28, 36, 46, 47])
(7.803498633820242, [2, 19, 19, 22, 22, 23, 25, 26, 36, 36])
(6.471209436512334, [2, 8, 19, 22, 23, 23, 28, 36, 47, 47])
(7.959323215038554, [6, 9, 9, 12, 22, 27, 27, 28, 47, 47])
(10.099957547750675, [9, 16, 19, 22, 26, 28, 28, 28, 42, 46])
(7.7064296959176675, [2, 2, 6, 27, 27, 27, 28, 36, 42, 47])
(6.679667036884902, [2, 8, 16, 22, 25, 36, 36, 42, 46, 47])
(14.012949153033894, [5, 6, 6, 6, 8, 19, 19, 22, 25, 28])
(10.345431411221046, [2, 6, 22, 22, 26, 28, 36, 36, 46, 46])
(6.6764278144264, [5, 5, 16, 19, 25, 26, 42, 42, 46, 47])
(5.966906107040374, [2, 2, 9, 23, 25, 26, 27, 42, 46, 47])
(4.984211477375441, [8, 12, 16, 16, 19, 22, 26, 28, 36, 46])
(4.231621906230883, [5, 6, 8, 9, 19, 23, 23, 26, 27, 28])
(11.176435874294155, [6, 6, 6, 8, 12, 28, 28, 46, 46, 47])
(7.113599927116694, [6, 8, 19, 25, 28, 36, 36, 42, 46, 47])
(5.403590740419912, [9, 16, 16, 19, 22, 25, 27, 36, 46, 47])
(5.550737893923385, [2, 8, 9, 19, 26, 26, 27, 42, 42, 46])
(7.915046637320256, [12, 16, 19, 22, 26, 27, 28, 42, 47, 47])
(6.552069338083239, [2, 6, 9, 16, 16, 23, 23, 25, 28, 47])
(11.006794501021009, [2, 5, 6, 8, 22, 25, 28, 36, 42, 47])
(11.559344133281037, [6, 6, 8, 9, 12, 19, 22, 25, 28, 42])
(5.292501690556316, [2, 6, 8, 9, 9, 19, 27, 36, 46, 47])
(7.055136113033312, [6, 6, 9, 9, 12, 28, 36, 46, 46, 47])
(5.517636502640172, [8, 9, 16, 16, 22, 25, 25, 26, 36, 46])
(7.603179139023158, [2, 6, 8, 9, 9, 22, 36, 36, 47, 47])
(4.3559646530306795, [6, 12, 16, 16, 19, 23, 26, 27, 28, 46])
(7.466221026630017, [2, 2, 6, 8, 12, 26, 26, 27, 42, 47])
(10.025395174498005, [16, 19, 22, 22, 22, 23, 23, 26, 42, 47])
(5.882666735772618, [5, 9, 12, 19, 22, 23, 25, 28, 36, 47])
(3.281140919505185, [6, 12, 16, 26, 26, 27, 27, 36, 46, 47])
(6.34072703859846, [5, 8, 9, 16, 19, 22, 26, 26, 42, 46])
(9.30222663803495, [6, 6, 8, 16, 22, 25, 26, 27, 36, 47])
(7.5633885243366645, [2, 8, 16, 22, 22, 23, 26, 36, 42, 46])
(4.955213179694436, [2, 5, 6, 8, 8, 8, 8, 9, 16, 46])
(7.731127394392328, [12, 22, 23, 26, 28, 28, 42, 46, 46, 47])
(4.778597842309573, [6, 8, 8, 8, 9, 16, 19, 23, 28, 46])
(8.313259757156144, [2, 5, 6, 6, 9, 19, 27, 27, 42, 46])
(12.600525948196815, [6, 19, 19, 22, 28, 47, 47, 47, 47, 47])
(5.207188748911839, [8, 12, 19, 22, 25, 26, 27, 36, 42, 46])
(5.567035964282161, [6, 16, 19, 23, 26, 36, 36, 42, 42, 42])
```

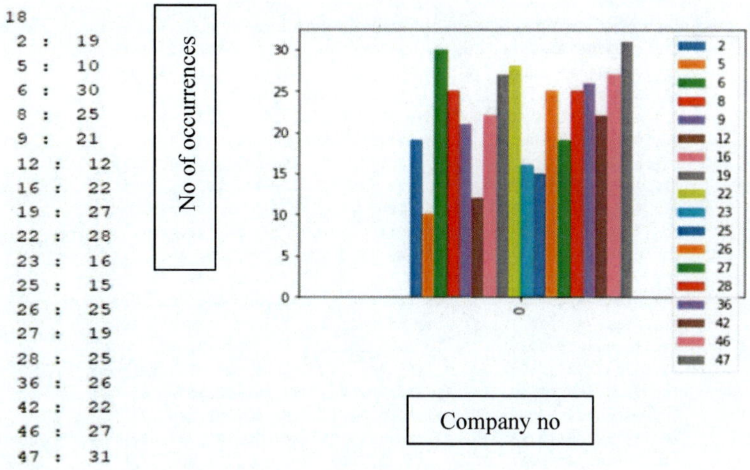

Fig. 2 Frequency of occurrence of each company in iteration-1

In the following Table 3, here the highest fitness is 9.883.

In Fig. 3, there is a graph stating which company has come how many time bases on the result of Iteration-3.

Iteration 4

Final Population after 1000 iterations (Fitness of chromosome [chromosome]).

In the following Table 4, here the highest fitness is 16.0653.

In Fig. 4, there is a graph stating which company has come how many time bases on the result of Iteration-4.

Iteration 5

Final Population after 1000 iterations (Fitness of chromosome [chromosome]).

In the following Table 5, here the highest fitness is 16.224.

In Fig. 5, there is a graph stating which company has come how many time bases on the result of Iteration-5 (Fig. 6).

In the following table, a comparative study has been analyzed. It represents the detail literary proposal related to the different algorithm that has been used in last few year about stock market price prediction using genetic algorithm. It can be observed from the table that the different research method has been used implementing genetic algorithm but, in this proposal, the concept of genetic algorithm has been incorporated with the nifty top 50 shares to evaluate the safe stocks in nifty top 50 shares rank wise (Table 6).

Table 2 Frequency of occurrence of each company in iteration-2

```
(15.422950670721594, [2, 4, 13, 14, 19, 19, 33, 33, 33, 41])
(14.337076939820571, [2, 19, 19, 19, 22, 26, 30, 33, 33, 40])
(6.959354677142147, [3, 3, 3, 14, 14, 15, 20, 22, 26, 39])
(12.40124851189224, [2, 3, 4, 13, 15, 19, 19, 30, 33, 33])
(11.8858267859594, [2, 3, 3, 13, 19, 20, 24, 28, 33, 33])
(14.307681384563542, [2, 3, 13, 14, 15, 15, 26, 30, 33, 33])
(11.762808306511074, [2, 3, 13, 14, 22, 22, 22, 26, 30, 39])
(11.404140728845814, [14, 19, 20, 26, 33, 33, 33, 39, 39, 39])
(7.79377421716392, [2, 2, 13, 14, 20, 20, 24, 33, 40, 40])
(9.08966747900847, [2, 2, 2, 4, 13, 14, 15, 26, 28, 30])
(12.92893108656512, [2, 2, 2, 13, 14, 22, 28, 28, 28, 41])
(12.213790916149337, [2, 4, 14, 24, 26, 33, 33, 39, 39, 40])
(13.299187532259737, [4, 13, 13, 32, 32, 33, 33, 33, 39, 40])
(12.901083875707531, [2, 13, 13, 14, 15, 15, 32, 33, 40, 40])
(9.435447558432534, [2, 2, 19, 20, 24, 26, 26, 28, 33, 40])
(4.032544900116557, [4, 4, 19, 19, 20, 24, 26, 32, 33, 39])
(10.350057235570091, [2, 4, 13, 13, 15, 19, 20, 22, 24, 33])
(6.399914866642963, [3, 13, 14, 26, 26, 26, 28, 28, 30, 32])
(10.93919276159381, [2, 4, 14, 26, 26, 30, 33, 33, 40, 41])
(8.708025641175553, [4, 13, 15, 15, 15, 24, 24, 32, 40, 40])
(11.402291048326472, [2, 2, 4, 14, 20, 22, 30, 32, 33, 33])
(6.727622205705746, [2, 4, 4, 15, 19, 22, 28, 30, 32, 39])
(7.795706528507566, [4, 24, 32, 33, 33, 39, 39, 40, 41, 41])
(8.364620914668707, [13, 13, 15, 15, 19, 22, 26, 26, 32, 39])
(5.315531210194195, [14, 14, 19, 24, 32, 32, 39, 40, 40, 41])
(1.340464719940357, [4, 13, 20, 20, 24, 24, 28, 28, 32, 32])
(3.375779159165104, [3, 3, 19, 22, 26, 32, 39, 39, 41, 41])
(7.838595943245449, [20, 30, 32, 33, 33, 39, 39, 40, 40, 40])
(6.385409485979251, [15, 19, 19, 19, 22, 28, 30, 39, 39, 41])
(9.909985683802482, [3, 4, 15, 24, 26, 33, 33, 39, 39, 41])
(6.594353937195651, [13, 14, 15, 15, 20, 24, 30, 33, 39, 41])
(9.779324185358538, [2, 2, 2, 3, 3, 4, 15, 22, 24, 32])
(9.310346456091867, [4, 13, 20, 22, 22, 22, 24, 26, 28, 32])
(0.3105353135659232З, [3, 13, 13, 19, 20, 20, 24, 32, 40, 41])
(6.594526923722892, [13, 14, 19, 24, 24, 26, 26, 28, 32, 39])
(12.393543386637623, [2, 4, 4, 13, 13, 14, 14, 15, 28, 33])
(9.716241294258383, [4, 14, 15, 19, 19, 19, 22, 24, 26, 40])
(4.9224911193093845, [20, 20, 20, 20, 24, 33, 33, 33, 40, 41])
(8.81484045029628, [2, 13, 19, 22, 26, 32, 32, 32, 33, 39])
(6.199118696641764, [13, 14, 15, 20, 24, 26, 28, 30, 40, 40])
```

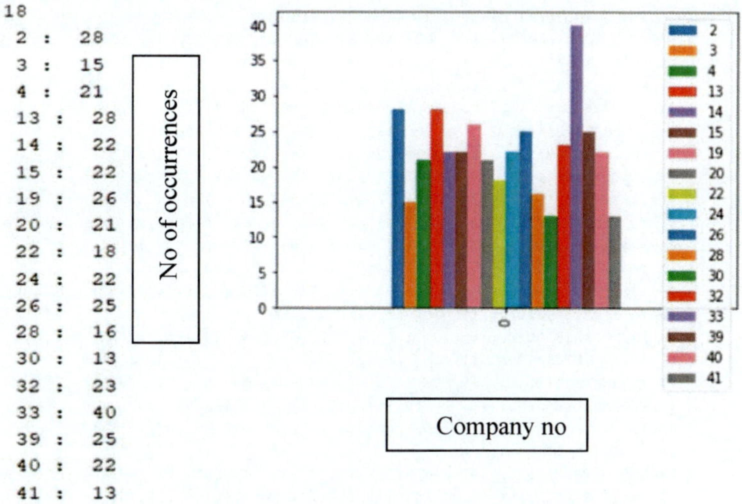

Fig. 3 Frequency of occurrence of each company in Iteration-2

5 Conclusion

Stock market investment is an essential component of financial planning to meet individual's financial goals. A major sum of domestic money is invested in stock market, which happens to be growing exponentially. Hence, stock market efficiency lies in its ability to give a predictable and striking return to investors. There are several local and global factors and events impact the stock market performance. Even if the stock market is unpredictable in a short-term standpoint, it shows a change in mean over longer time horizon and that is possibly how people are making money by investing.

It can be seen from the above outputs that it has been possible to scale down at least 15–20 companies from the 50 which are more or less safe to invest in. The solution obtained is not optimal as it has been restricted by the data which has been collected. Once more data can be collected, it can be done at a more rigid and optimal result and concluded which Company will be the safest to invest in. Fitness function also needs to be improved to get to a better solution.

Table 3 Frequency of occurrence of each company in iteration-3

```
(-4.357032639600597, [12, 23, 30, 34, 36, 36, 41, 45, 47, 50])
(1.297039566711219, [16, 24, 24, 30, 34, 36, 41, 42, 44, 45])
(4.438011338978527, [12, 12, 12, 25, 30, 34, 41, 42, 44, 44])
(9.095822629747214, [1, 18, 34, 36, 42, 44, 44, 45, 50, 50])
(-2.0828899681438102, [9, 10, 23, 24, 30, 34, 41, 44, 45, 45])
(5.640317842831787, [9, 9, 18, 25, 30, 34, 44, 47, 47, 50])
(8.610807643685673, [9, 9, 23, 24, 36, 41, 41, 44, 44, 50])
(1.1331318262475403, [1, 10, 16, 18, 30, 41, 42, 45, 45, 50])
(2.137401043692903, [16, 23, 34, 34, 36, 42, 44, 47, 50, 50])
(2.104524604531627, [1, 12, 18, 23, 30, 34, 44, 45, 47, 50])
(2.443813626013594, [9, 10, 12, 16, 18, 23, 24, 25, 36, 50])
(-1.3605223722631816, [1, 18, 23, 25, 34, 36, 36, 36, 47, 47])
(1.7475793264311017, [9, 10, 16, 16, 23, 25, 25, 30, 36, 42])
(8.293750384840532, [1, 10, 18, 24, 41, 42, 44, 44, 45, 45])
(2.3016911576538095, [1, 16, 25, 25, 25, 25, 34, 41, 45, 50])
(7.3772687249394115, [9, 12, 16, 18, 24, 41, 44, 45, 50, 50])
(8.669245953698413, [1, 1, 10, 12, 12, 24, 25, 42, 44, 50])
(6.09819059241464, [1, 1, 9, 10, 12, 18, 23, 44, 47, 47])
(-1.006210985617903, [9, 10, 12, 16, 16, 23, 36, 41, 45, 50])
(4.575304736318573, [10, 16, 18, 24, 24, 36, 44, 45, 45, 47])
(3.6441866482325764, [18, 18, 18, 18, 24, 30, 34, 44, 45, 50])
(-4.327743165609461, [1, 9, 12, 16, 23, 23, 34, 41, 45, 47])
(1.6199426705469693, [9, 10, 10, 12, 18, 36, 45, 47, 47, 50])
(3.3067476891360004, [1, 10, 16, 24, 25, 25, 30, 36, 45, 50])
(7.646526782835629, [1, 9, 9, 10, 12, 16, 18, 44, 44, 45])
(-3.520157525981196, [12, 18, 18, 18, 23, 30, 34, 34, 41, 44])
(3.3712140834424775, [1, 9, 9, 10, 12, 16, 42, 45, 47, 50])
(0.4002567260386709, [1, 9, 10, 10, 18, 24, 36, 41, 41, 47])
(-1.1369176590787102, [1, 9, 18, 25, 34, 36, 45, 45, 47, 47])
(4.818359052895983, [10, 10, 18, 23, 23, 25, 41, 42, 44, 47])
(9.883633983310855, [1, 16, 18, 25, 34, 42, 44, 44, 47, 47])
(-1.298440539916259, [10, 10, 12, 12, 16, 36, 36, 42, 45, 47])
(7.124729251959674, [9, 16, 18, 23, 34, 41, 42, 44, 44, 50])
(-1.2200643898875387, [1, 9, 12, 16, 23, 36, 41, 41, 42, 45])
(3.9855285265270775, [9, 9, 12, 18, 24, 25, 30, 36, 47, 50])
(-11.70353888412442, [10, 10, 18, 18, 23, 30, 30, 34, 34, 34])
(5.211803972218502, [9, 10, 10, 16, 23, 24, 25, 47, 50, 50])
(-0.4771368349289751, [1, 10, 16, 18, 18, 36, 41, 45, 45, 47])
(3.727106795515792, [9, 9, 16, 16, 30, 36, 36, 41, 42, 44])
(-6.300756432511537, [1, 9, 23, 24, 25, 30, 34, 34, 36, 45])
```

It needs to be improved the data collection strategy so that we can run the algorithm for a greater number of iterations and hence arrive at a more accurate and optimal result. It needs to improve the fitness function by introducing a reduction factor for companies going out of the Nifty50 so that selection and crossover processes are improved thereby increasing the accuracy of the final result. Moreover, nearly after every six months the Nifty 50 shares are revised.

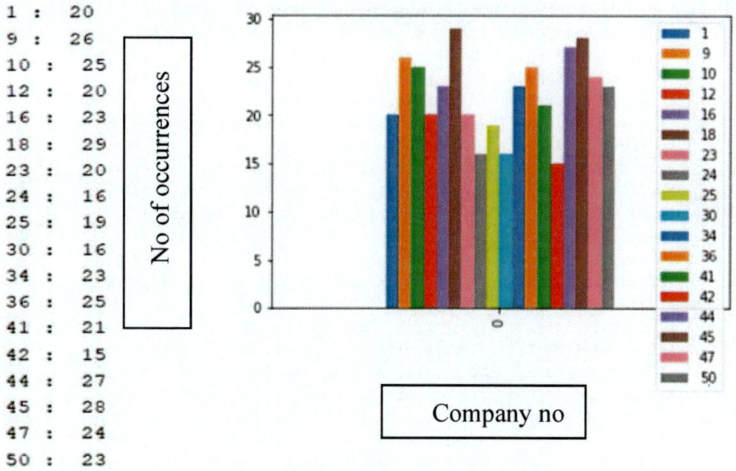

Fig. 4 Frequency of occurrence of each company in iteration-3

Table 4 Frequency of occurrence of each company in iteration-4

```
(9.766301406753147, [5, 5, 5, 28, 43, 43, 45, 45, 49, 49])
(8.167271231568817, [5, 15, 15, 15, 21, 21, 29, 40, 45, 49])
(4.071122563673489, [2, 2, 10, 20, 23, 29, 40, 40, 43, 45])
(7.760931565190555, [2, 15, 15, 17, 23, 28, 29, 29, 35, 45])
(3.8166893491079, [5, 15, 17, 23, 23, 23, 28, 29, 35, 35])
(10.406738759125568, [2, 17, 20, 20, 40, 40, 49, 49, 49, 49])
(9.401386742378937, [2, 17, 20, 21, 29, 29, 29, 42, 43, 49])
(11.307571901996132, [2, 2, 2, 17, 29, 29, 40, 42, 45, 49])
(6.907823271659703, [5, 15, 17, 20, 21, 28, 43, 43, 45, 48])
(7.582213548106177, [2, 17, 20, 21, 35, 40, 43, 43, 45, 49])
(6.446313351950597, [17, 17, 17, 20, 23, 23, 40, 40, 42, 42])
(12.267574173797998, [2, 2, 2, 5, 15, 15, 28, 40, 42, 42])
(11.725537490881997, [5, 15, 15, 17, 21, 35, 43, 48, 49, 49])
(7.686502468744374, [10, 15, 20, 21, 28, 29, 35, 43, 43, 49])
(14.673604422549193, [5, 15, 15, 15, 28, 29, 29, 43, 49, 49])
(5.08262256284374, [5, 17, 21, 23, 35, 35, 42, 42, 42, 48])
(5.336911093876871, [10, 15, 15, 20, 21, 23, 29, 40, 42, 49])
(6.890525988673131, [10, 10, 17, 17, 23, 28, 29, 42, 45, 49])
(16.065306346468766, [15, 17, 17, 28, 28, 28, 40, 40, 49, 49])
(8.669692083581097, [17, 17, 20, 21, 28, 29, 40, 42, 43, 48])
(7.897247206591281, [2, 10, 23, 28, 28, 28, 40, 45, 48, 49])
(4.41025723953854, [5, 10, 21, 23, 29, 40, 42, 45, 48, 49])
(3.2422211962165424, [5, 17, 20, 20, 21, 28, 28, 35, 35, 49])
(5.238576869012938, [5, 23, 23, 23, 28, 28, 40, 42, 43, 45])
(8.488561580318583, [15, 17, 21, 23, 35, 42, 43, 43, 48, 48])
(1.7778337492551737, [2, 10, 21, 23, 23, 29, 40, 45, 48, 48])
(0.4329644046016032, [10, 15, 21, 45, 45, 45, 48, 48, 48, 48])
(0.7244808382705041, [2, 20, 20, 20, 21, 28, 29, 40, 45, 49])
(1.663476462892655, [10, 15, 20, 20, 21, 23, 28, 28, 29, 40])
(-0.9389938050717848, [2, 5, 10, 20, 20, 23, 23, 29, 35, 43])
(0.4663500671167291, [5, 10, 15, 20, 29, 42, 42, 45, 45, 45])
(7.298382817128207, [2, 10, 10, 21, 21, 28, 29, 40, 42, 49])
(-0.2237512422959771, [10, 15, 17, 20, 20, 21, 28, 29, 45, 45]
(6.182800430898631, [5, 15, 15, 17, 20, 23, 29, 29, 35, 43])
(5.401496911376235, [2, 5, 10, 15, 20, 21, 40, 40, 43, 48])
(1.436386086808355, [2, 5, 20, 20, 21, 23, 28, 28, 35, 42])
(0.038223228910513996, [5, 10, 10, 17, 17, 20, 20, 35, 43, 45]
(1.1373786121605287, [2, 10, 20, 20, 20, 40, 42, 42, 43, 48])
(12.394737068900382, [10, 17, 17, 21, 42, 42, 48, 49, 49, 49])
(3.828505378895589, [2, 10, 20, 21, 29, 29, 40, 45, 48, 49])
```

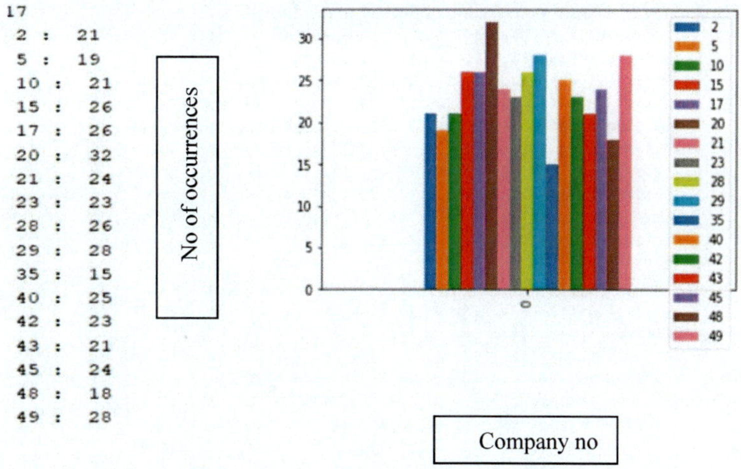

Fig. 5 Frequency of occurrence of each company in iteration 4

Table 5 Frequency of occurrence of each company in iteration-5

```
(13.250888136760063, [6, 16, 22, 30, 41, 42, 44, 44, 45, 47])
(3.948902470556181, [1, 1, 12, 22, 24, 24, 29, 30, 45, 45])
(5.624616406397133, [6, 12, 16, 27, 27, 27, 30, 41, 42, 44])
(6.744741389114329, [6, 12, 12, 16, 22, 26, 29, 45, 47, 47])
(10.592827179006576, [1, 6, 16, 22, 26, 26, 30, 41, 42, 44])
(6.7498284574324305, [1, 22, 24, 26, 30, 41, 44, 45, 45, 47])
(14.993751313492687, [1, 6, 6, 22, 22, 41, 44, 45, 47, 47])
(9.299946665876334, [16, 16, 22, 26, 41, 42, 42, 44, 45, 47])
(13.791218065444436, [6, 12, 16, 22, 24, 27, 41, 42, 44, 44])
(16.224447569552787, [6, 6, 6, 22, 22, 22, 24, 27, 35, 47])
(1.963990148135593, [1, 1, 1, 12, 12, 16, 29, 41, 41, 47])
(12.857641672640305, [1, 22, 29, 35, 41, 41, 42, 44, 44, 47])
(11.579553663210383, [1, 6, 11, 11, 16, 24, 24, 26, 26, 44])
(11.318714004688339, [6, 6, 11, 11, 16, 22, 27, 27, 29, 42])
(6.329747823220915, [1, 12, 24, 26, 29, 35, 35, 42, 44, 45])
(7.721717142436868, [1, 6, 11, 24, 24, 29, 29, 29, 35, 41])
(4.933878598029603, [11, 12, 26, 26, 27, 29, 30, 30, 35, 44])
(10.266989162538229, [1, 6, 6, 11, 12, 16, 22, 24, 27, 47])
(9.009472146321297, [6, 16, 27, 30, 30, 41, 41, 44, 44, 47])
(8.515400971994962, [1, 6, 16, 27, 35, 41, 44, 44, 45, 45])
(1.5553100275380556, [1, 6, 11, 12, 12, 26, 30, 35, 41, 45])
(2.3010140511753825, [22, 24, 26, 26, 27, 27, 30, 35, 35, 45])
(5.910211801454566, [6, 11, 16, 16, 26, 29, 41, 42, 45, 47])
(5.422741863949067, [1, 12, 22, 24, 24, 26, 29, 29, 41, 45])
(7.372351170053568, [1, 1, 29, 30, 42, 42, 44, 45, 45, 47])
(6.655280400574467, [6, 6, 11, 26, 27, 27, 30, 41, 47, 47])
(12.145300564257347, [1, 1, 11, 11, 11, 22, 24, 30, 44, 47])
(7.422437026882567, [12, 16, 16, 22, 24, 27, 29, 29, 42, 42])
(3.9374876344790044, [6, 12, 27, 29, 41, 41, 42, 45, 47, 47])
(9.731559270193541, [12, 12, 22, 22, 24, 26, 27, 29, 41, 44])
(7.49668689336704, [6, 11, 27, 27, 29, 29, 29, 30, 42, 47])
(6.882241621044459, [12, 12, 16, 22, 22, 22, 26, 26, 41, 41])
(8.42916257690907, [6, 11, 22, 26, 26, 27, 27, 29, 35, 42])
(6.81134110108756, [1, 1, 6, 16, 24, 26, 26, 26, 27, 42])
(9.71881415144381, [1, 1, 6, 6, 11, 22, 24, 26, 30, 41])
(9.21561315538191, [6, 6, 6, 16, 16, 24, 29, 30, 41, 42])
(6.945908537014381, [6, 6, 6, 12, 24, 27, 29, 30, 30, 35])
(7.889150215431651, [1, 22, 22, 27, 27, 29, 30, 35, 42, 47])
(12.071881347166968, [1, 6, 6, 11, 12, 12, 24, 29, 44, 47])
(6.649175438571645, [6, 11, 12, 22, 27, 27, 35, 41, 42, 47])
```

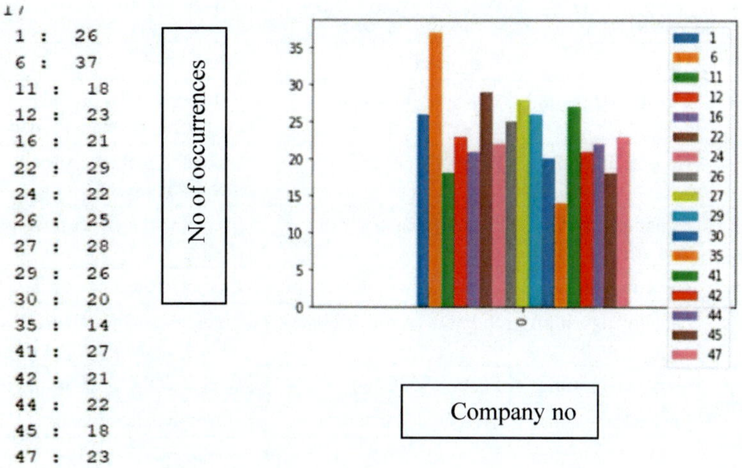

Fig. 6 Frequency of occurrence of each company in iteration 5

Table 6 Comparative summary report

Proposals	Parameter 1	Parameter 2	Summary
[1]	Genetic algorithm	Fuzzification	Reduced the dimensionality of the feature space and removed irrelevant factors involved in stock price prediction
[2]	Genetic algorithm	Artificial Neural Network	Reduced the complexity and dimensionality of the feature space and enhanced the generalizability of the classifier
[3]	Historical Data	Mathematical calculation	Established relationships between stock price changes on Mondays and Fridays in the stock market
[4]	Genetic algorithm	Regression method	Predicted sediment load and discharge
[5]	Genetic algorithm	Case-based reasoning	Predicted financial bankruptcy
[6]	Genetic algorithm	Technical Indexes (RSI)	Optimized the timing of an automated trader
[7]	Genetic algorithm	Technical index (DAX30)	Generated the buying and selling signals of 30 stocks without devise any criterion of profit cashing and loss cutting
[8]	Genetic algorithm	Financial news, Technical indicator	Optimized the stock prediction

(continued)

Table 6 (continued)

Proposals	Parameter 1	Parameter 2	Summary
[9]	Genetic algorithm	k-means clustering	Enhanced stock prediction analyzing twitter sentiment
[10]	Genetic algorithm	Financial news	Predicted market return by stochastic optimization
[11]	Genetic algorithm	Target price of script	Predicted the target price of stocks
[12]	Genetic algorithm	Artificial neural network	Predicted the stock price
[13]	Genetic algorithm	Support vector machine	Predicted the stock price by collecting data using web robot
[14]	Genetic algorithm	Feature selection	Predicted the stock price trends
[15]	Genetic algorithm	Weighted F-measure	Object localization
[16]	Genetic algorithm	Ibo vespa index	Predicted of weekly prices movement
[17]	Genetic algorithm	Fuzzy logic	Optimized the interval and predicted a particular share "closing price"
[18]	Genetic algorithm	Ibo vespa index	Designed a decision system to "Buy" or "Sell" stocks in Sao Paulo Stock Exchange
[19]	Genetic algorithm	Historical finance dataset	Designed a system which will prompt for correcting the cut-off value for stop-loss to the ORB intraday policy
[20]	Genetic algorithm	Fuzzy logic	Improved the correctness of the prediction of stocks
[21]	Genetic algorithm	Support vector machine	Sentiment analyzed using twitter data then optimized the trading rule to increase profit
[22]	Genetic algorithm	Neural Network	Forecasted "closing price" of rare earth shares using historical data of the same
[23]	Genetic algorithm	RSI	Forecasted the price of individual share
Proposed	Genetic algorithm	Nifty top 50 shares	Prediction of safe stocks rank wise from nifty top 50 shares

References

1. Kyoung-jae Kim WBL (2004) Stock market prediction using artificial neural networks. In Neural computing and applications, pp 255–260
2. Kyoung-jae Kim IH (2000) Genetic algorithms approach to feature discretization in artificial neural networks for the prediction of stock price index. In Expert systems with applications, pp 125–132
3. Cross F (1973) The behavior of stock prices on Fridays and Mondays. Financial Analysts J 29(6):67–69. https://doi.org/10.2469/faj.v29.n6.67
4. Altunkaynak A (2009) Sediment load prediction by genetic algorithms. In Advances in Engineering Software, pp 928–934
5. Hyunchul Ahn K (2009) Bankruptcy prediction modeling with hybrid case-based reasoning and genetic algorithms approach. Applied Soft Computing, 599–607
6. David de la Fuente AG (2006) Genetic algorithms to optimize the time to make stock market investment. GECCO, pp 1857–1858
7. Cyril Schoreels BL (2004) Agent based genetic algorithm employing financial Technical analysis for making trading decisions using historical equity market data. In IEEE Computer Society Washington, DC, USA, pp 421–424
8. Chen C, Shih P (2019) A stock trend prediction approach based on Chinese news and technical indicator using genetic algorithms. In 2019 IEEE Congress on Evolutionary Computation (CEC), Wellington, New Zealand, pp 1468–1472
9. Desokey EN, Badr A, Hegazy AF (2017)Enhancing stock prediction clustering using K-means with genetic algorithm. In 2017 13th International Computer Engineering Conference (ICENCO), Cairo, Egypt, pp 256-261
10. Fu X, Ren X, Mengshoel OJ, Wu X (2018) Stochastic optimization for market return prediction using financial knowledge graph. In 2018 IEEE International Conference on Big Knowledge (ICBK), Singapore, pp 25–32
11. Sable S, Porwal A, Singh U (2017)Stock price prediction using genetic algorithms and evolution strategies. In 2017 International conference of Electronics, Communication and Aerospace Technology (ICECA), Coimbatore, Tamil Nadu, India, pp 549–553
12. Solin MM, Alamsyah A, Rikumahu B, Arya Saputra MA (2019) Forecasting portfolio optimization using artificial neural network and genetic algorithm. In 2019 7th International Conference on Information and Communication Technology (ICoICT), Kuala Lumpur Malaysia, pp 1–7
13. Wang C-T, Lin Y-Y (2015) The prediction system for data analysis of stock market by using Genetic Algorithm. In 2015 12th International Conference on Fuzzy Systems and Knowledge Discovery (FSKD), Zhangjiajie, China, pp 1721–1725
14. Xia T, et al. (2018)Improving the performance of stock trend prediction by applying GA to feature selection. In 2018 IEEE 8th International Symposium on Cloud and Service Computing (SC2), Cairo, Egypt, pp 122–126
15. Jagtap VA (2012) Genetic programming for object localization using fitness function based on relative localization weighted F-measure. IJCET, 659–666
16. Gonzalez RT, Padilha CA, Barone DAC (2015)Ensemble system based on genetic algorithm for stock market forecasting. In 2015 IEEE Congress on Evolutionary Computation (CEC), Sendai, Japan, pp 3102–3108
17. Watada J, Zhao J, Matsumoto Y (2016) A genetic rough set approach to fuzzy time-series prediction. In 2016 Third International Conference on Computing Measurement Control and Sensor Network (CMCSN), Matsue, Japan, pp 20–23.
18. Nascimento TP, Labidi S, Neto PB, Timbó N, Almeida A (2015) A system based on genetic algorithms as a decision making support for the purchase and sale of assets at SÃ£o Paulo Stock Exchange. In International Conference on Computer Vision and Image Analysis Applications, pp 1–6
19. Syu J-H, Wu M-E, Chen C-H, Ho J-M (2020) Threshold-adjusted ORB strategies with genetic algorithm and protective closing strategy on Taiwan futures market. In ICASSP 2020 IEEE

International Conference on Acoustics, Speech and Signal Processing (ICASSP), Brighton, UK, pp 1778–1782
20. Sachdev A, Sharma V (2015) Stock forecasting model based on combined Fuzzy time series and genetic algorithm. In 2015 International Conference on Computational Intelligence and Communication Networks (CICN), Jabalpur, pp 1303–1307
21. Simões C, Neves R, Horta N (2017)Using sentiment from Twitter optimized by Genetic Algorithms to predict the stock market. In 2017 IEEE Congress on Evolutionary Computation (CEC), San Sebastián, Spain, pp 1303–1310
22. Zhang H, Sun R (2016) Parameter analysis of hybrid intelligent model for the prediction of rare earth stock futures. In 2016 12th International Conference on Natural Computation, Fuzzy Systems and Knowledge Discovery (ICNC-FSKD), Changsha, China, pp 835–839
23. Gite B, Sayed K, Mutha N, Marpadge S, Patil K (2017)Surveying various genetic programming (GP) approaches to forecast real-time trends & prices in the stock market. In 2017 Computing Conference, London, UK, pp 131–134
24. Goldberg DE (1989) Genetic Algorithms in Search, Optimization and Machine Learning. Addison-Wesley Longman Publishing Co, Boston, MA, USA
25. Davis L (1991) Handbook of genetic algorithms (1st ed, Van Nostrand Reinhold (ed)). Computer Aided Architectural Design, New York

Chapter 10
Urban Growth Prediction of Kolkata City Using SLEUTH Model

Krishan Kundu, Prasun Halder, and Jyotsna Kumar Mandal

Abstract Due to increasing urbanization, urban growth or sprawl monitoring and measurement is needed in the developing countries like in India. For findings the urban growth and prediction of upcoming possible development of Kolkata city the SLEUTH model is employed. It is one of the most important urban development models which is utilized all over the world. The model was calibrated through the historical past data which are extracted from the satellite images in different time period. Six input data are used in this model such as slope, land use, excluded, urban extent, transportation, and hillshade. All the inputs were derived from the satellite image, and that was classified through the Maximum Likelihood classification technique. In this research, five Landsats temporal data (1978, 1988, 2000, 2010 and 2020) were used for prediction of urban growth of Kolkata City. The historical urban scenario is presented in this study which allowed urban expansion persistence in the previous trend. Calibration results show that the spread coefficient value is high which indicates the future prediction of Kolkata city is edge enlargement. Study revealed that in future more urban expansion may happen from 2020 to 2040 in the north-east and south-east positions of the Kolkata City. Besides, it is also observed that in future in the year of 2040 about 70% of total study area may be occupied by the urban area.

Keywords Multi-temporal data · Kolkata city · Supervised classification · Urban expansion · SLEUTH model

K. Kundu (✉)
Department of Computer Science and Engineering, Government College of Engineering and Textile Technology, Serampore, Hooghly, West Bengal, India

P. Halder
Department of Computer Science and Engineering, Ramkrishna Mahato Government Engineering College, Purulia, West Bengal, India

J. K. Mandal
Department of Computer Science and Engineering, University of Kalyani, Kalyani, Nadia, West Bengal, India

J. K. Mandal et al. (eds.), *Proceedings of International Conference on Innovations in Software Architecture and Computational Systems*, Studies in Autonomic, Data-driven and Industrial Computing, https://doi.org/10.1007/978-981-16-4301-9_11

1 Introduction

With the perspective of global scenario, particularly rising countries like India have practiced an uncontrolled urban expansion due to its enlargement of residents, industrialization and monetary activities in recent past [1]. It is definitely associated with economic development, although fast and unintended urbanization involves in fall down of natural and social consistency which may be harmful for the environment [2]. A hundred years before, about 15% of the total residents were livelihood in the urban areas, although in the present scenario proportion is almost 50%. In the year of 2030, around 5 billion communities are projected to living in towns which lead to 60% of total population [3]. Therefore, a well-organized and structure planned urbanization may lead to the better way of development like economic activities, education facilities, provide jobs, good healthcare services, and so on. Hence, there is need a tool for Kolkata planers which controlling the urban planning, monitoring, and calculating the future possible extension that the cities become better and more sustainable. In this view, for investigation and prediction of dynamic urban expansion have been implemented using the computerized models which plays an important role [4, 5]. Today, vast development in the field of computer science, geographic information systems, and remote sensing technology has permitted to make more spatial modeling technologies such as neural networks, multi-agent models, and cellular automata. These modeling technologies are enormously employed in the area of urban development. Policy makers use these spatial models for investigation of different scenarios of urban extension, land use land cover changes, etc. [6, 7].

SLEUTH model widely employed in the modeling of land use land cover alternation. For betterment of modeling, this model incorporates the concept of cellular automata (CA) [8]. This model forecast the upcoming probable extension of urban area with respect to previous records under the various circumstances. The naming convention of SLEUTH model comes from the name of the inputs where the inputs are needed to execute the model. The inputs are slope, land use, elevation, urban, transport, and hillshade [9, 10]. This model extensively applied to simulate the urban expansion of numerous cities throughout the Earth. This model is fruitfully executed in various cities in USA such as Chicago, south coast California, San Francisco, Chesapeake, and Washington [11]. In the Asia region, this model is also implementing in various countries like Chiang Mai in Thailand, and Taipei in Taiwan [12]. While in south Asia only one city (Pune in India) is effectively applied in this model for prediction of urban extension. Till now no one has studied the urban growth model of Kolkata City using SLEUTH model [13]. With the help of remote sensing data and SLEUTH model may fulfill the gap for prediction of urban extension of Kolkata city. The main intention in this study is to test and simulating the urban growth of Kolkata city from 1975 to 2020 with the help of old statistics records. Moreover, it is also finding the prediction of possible expansion of Kolkata city from 2020 to 2040.

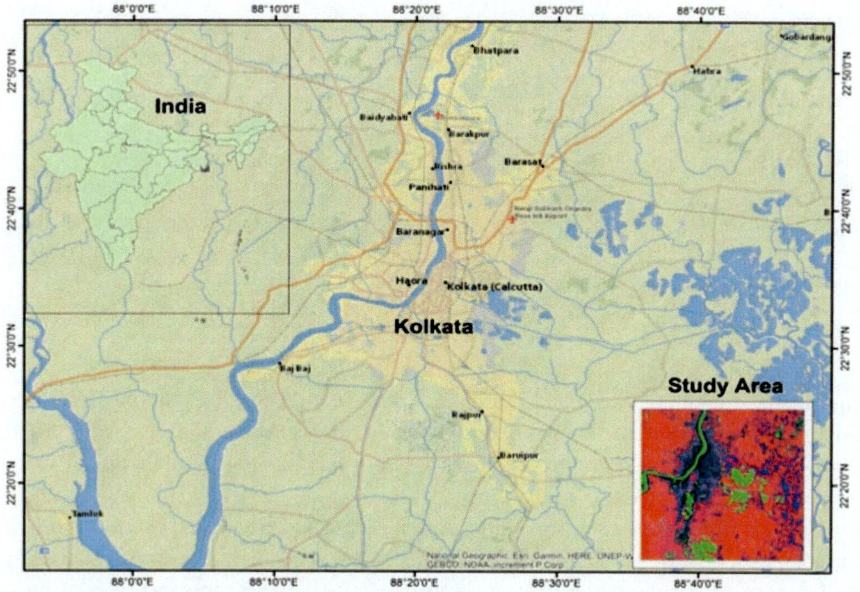

Fig. 1 Geographical place of Kolkata city

2 Material and Methods

2.1 Study Area

Present study area (Kolkata) is positioned in the state of west Bengal in India which extent from 22° 06′ 12″ N to 22° 45′ 41″ N latitude and from 88° 10′ 17″ E to 88° 31′ 11″ E longitude (Fig. 1). The city is placed besides the River Ganges and 154 km distances from the Bay of Bengal. This region is the economic corridor, commercial and business hub in the sates of eastern and north-eastern India. The city serves as the capital of India in the British Governance until 1911 and it is more than 300 years elderly. It is one of the biggest metropolitan cities in the world. Moreover, Kolkata urban area is continuously expanding from day by day due to quickly growth of population. Alongside the southern–northern of the River Hooghly is more extends in contrast to the other positions.

2.2 SLEUTH Model

SLEUTH is a Cellular Automata-based model (CA). This model consists of two sub-modules such as Urban Growth Model (UGM) and Deltatron Land Use Model (DLM). In SLEUTH model includes the transition rules of cellular automata. The

Table 1 Growth rules and model coefficient correlation in SLEUTH model

Type of growth	Growth coefficient	Descriptions
Spontaneous growth	Dispersion, Slope	Arbitrary translation from non-urban to urban cell
New spreading growth	Breed, Slope	Findings the new spreading centers from automatic growth
Edge growth	Spread, Slope	Extension of edge from spreading center
Road influenced growth	Breed, Road gravity, Slope, Spread	Extension of urbanization from both sides of road

input requirements of SLEUTH model are the layers describing land cover, slope, existing urbanized areas, excluded areas, and road networks. The dynamic urban growth in SLEUTH is composed of four growth rules like (1) spontaneous growth, (2) new spreading center, (3) Edge growth, and (4) road influenced growth. In order to each rule is employed during calibration phases, and is managed by through five parameters such as (1) dispersion, (2) breed, (3) spread, (4) slope resistance, and (5) road gravity. Every parameter value is set from 0 to 100. In spontaneous urban growth, every pixel has capability to translate the urban state (random urbanization) and is managed by the diffusion parameter. The new spreading center growth finds the new automatic urban pixels which become new urban spreading centers and is managed by breed parameter. Edge growth is simply occurred on the existing development of the urban areas. It is managed by the spread parameter; here, non-urban cell is influencing at least three near pixels which will become urbanized. The parameter of road influence is affected by the surroundings of road transportation networks and is managed by the diffusion, spread, and breed. The correlation between the five growth coefficients and four growth rules is presented in Table 1.

The circulation of overall external dispersive nature is estimated using the diffusion parameter. Newly creation of isolate settlement area is determined by the breed parameter. How much amount of scattering extension is happed from the existing settlement area is measured by spread parameter. The slope disagreement factor impacts the possibility of settlement extending up steeper slopes. New settlement along and near the road networks assess by the road gravity parameter. Generally, five steps are needed to implement the SLEUTH model of Kolkata City namely as compilation phase, creation of input data, model calibration, model prediction, and produce outputs. Here, Linux operating system is used to run this model.

2.3 Construction of Input Data

SLEUTH model inputs are two land use, four urban extents, two transportations, and one for each slope, exclusion, hillshade data. These data are required to findings the best-fit calibration outcomes which were derived from satellite images.

Table 2 Characteristics the input data

Input data	Data sources	Type of data and years
Land use	Image classification	Raster, 1978, 1988, 2000, 2010, 2020
Urban	Image classification	Raster, 1978, 1988, 2000, 2010, 2020
Transport	Digitize the image	Conversion from raster to vector, 1978, 2020
Slope	Generate digital elevation model	Raster
Hillshade	Generate digital elevation model	Raster
Excluded	Manual digitization	Conversion from raster to vector data

In this study, five Landsat satellite images (Multispectral sensor-1978, Thematic mapper-1988, Enhanced thematic mapper-2000, Enhanced thematic mapper-2010, and Operational land imager-2020) were used which have been acquired from United States Geological Survey (USGS). Collections of all the images are almost in the same season. If image contains any distortion or noise that can be minimized using geometric correction and radiometric calibration operation was performed with the help of GIS-based software. The images were classified using Maximum Likelihood classification (MLC) technique with the help of the training set. Land use layer was prepared using classification of satellite images. The urban layer was created from reclassify the land use land cover maps. Transport or road layer was derived from manual digitization of each satellite image. Slope layer was generated from Digital Elevation Model (DEM). The DEM was collected from the USGS. In this layer is allocated to percentage in slope changed to percentage slope. The hillshade layer was generated from the same DEM. The excluded layer contains some restricted areas where in future urban development is not possible such as water body, park, and green areas. All input layers data were prepared in the gif format. The data for input layers of the SLEUTH model is presented in Table 2. Figure 2 displayed the methodology used by present study area. Input layers of SLEUTH model data are represented in Fig. 3 (land use layer), Fig. 4 (urban layer), and in Fig. 5 (road or transport, hillshade, slope and excluded layers).

2.4 Model Calibration

To finds the appropriate values of five control variables such as coefficient of diffusion, breed, spread, slope resistance, and road gravity. Five calibration coefficients are incorporated together and their values starting from 0 to ending 100. The maximum value is set to 100 for each of diffusion, breed, spread, slope resistance and road gravity. Brute Force calibration process is used in SLEUTH model. The calibration process is carried out in three phases: coarse calibration, fine calibration, final calibration. For each phase, coefficient range, increment size and resolution of the input layers were changed. In other words, as the parameter range was narrowed down

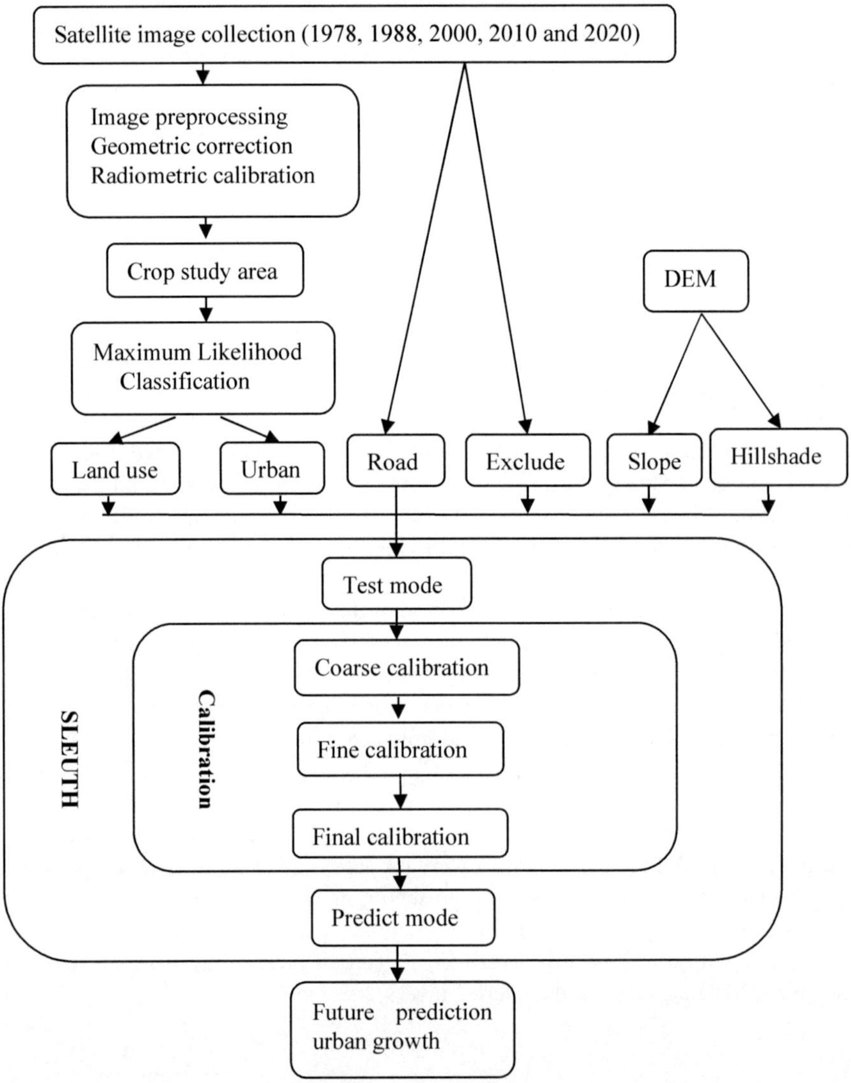

Fig. 2 Methodology used by present work

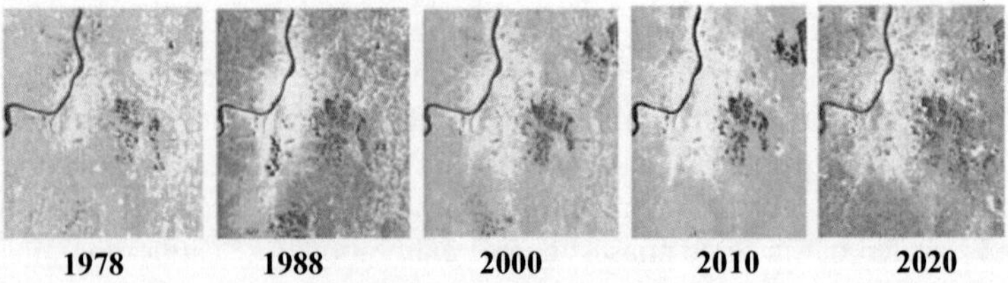

Fig. 3 Land use layers of 1978, 1988, 2000, 2010, and 2020

Fig. 4 Urban layers of 1978, 1988, 2000, 2010, and 2020

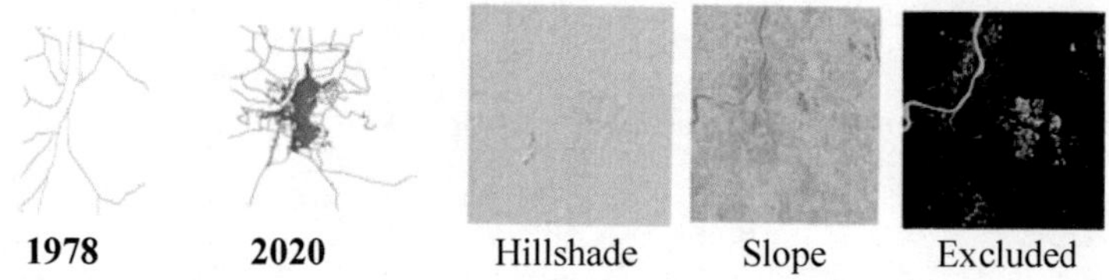

Fig. 5 Road or transport (1978, 2020), hillshade, slope and excluded layers

in each phase, the increment size became smaller and the resolution of the input images became better and closer to the full image resolution as in final calibration phase, the increment size was the lowest and the full resolution images were used as input layers. In this way, the number of steps was reduced, but searched the full range of solutions to obtain the suitable values. Finally, execute the calibration model and generates the thirteen least square regressions metric parameters are compare, r^2 population, edge r^2, R^2 cluster, Leesalee, average slope r^2, %urban, X_r^2, Y_r^2, and radius. Simulated and actual growth parameter is compared for the control years which are represented by the tabular structure.

3 Results and Discussion

3.1 Model Calibration Outcomes

For change detection of Kolkata city with respect to time there are five key attributes are used such as dispersion, breed, spread, slope resistance, and road gravity which are mainly controlled the urban expansion. Increasing of Breed attribute and shrink of slope resistance were significantly reflected on the urban changes during the calibration process. The increase of breed coefficient means finding the new urban centers and ultimately increased the urban area. Slope resistance coefficient value is decline means more structure was made as compared with the past structure in the particular slope region. A Leesallee shape index is used to measure the performance of the

Table 3 Best results for three phases of model calibration

Parameter/phase	Coarse	Fine	Final
Compare	0.954	0.931	0.911
r^2 population	0.987	0.983	0.982
Edge r^2	0.879	0.881	0.875
R^2 cluster	0.742	0.713	0.679
Leesalee	0.587	0.589	0.591
Average slope r^2	0.876	0.891	0.899
%Urban	0.931	0.971	0.989
X_r^2	0.890	0.898	0.891
Y_r^2	0.879	0.881	0.889
Radius	0.87	0.85	0.86
Diffusion	20	33	25
Breed	10	45	65
Spread	45	70	92
Slope	30	55	45
Road gravity	20	25	47

spatial fit which is the ratio between the subtraction and the combination of the simulated and actual urban areas in SLEUTH model. The value of 1 represents the perfect spatial matching. After final calibration Leesallee index value 0.591 was generated.

The best-fit calibration result of Kolkata City is presented in Table 3. The breed coefficient is relatively high that means a new spreading center was found that lead to an urban growth and its value was reported as 65 after final calibration. A relatively high spread coefficient (92) showed that organic or edge growth was found in the new spreading center or the existing urban area. For urban growth topographic structure is not limited factor whenever the slope coefficient value is relatively medium (45) which are verified by some field observation data. Diffusion coefficient value is relatively low (25) which means that urban expansion of Kolkata city is relatively compact and more urbanization happened which is closer to the existing urban area and new urban centers. Road gravity coefficient value is comparatively medium (47) which indicate that major development is occurred along the main transportation network such as road highways or railways. Figure 6 illustrated the highest value of urban coefficient is spread, and then followed by breed, road gravity, slope resistance, and dispersion.

3.2 Urban Growth Prediction by 2040

SLEUTH model finds the future projection of urban growth by 2040 with the help of calibration data which is presented in Table 3. The initial seed layer was selected as 2000 and determines the historical urban scenario of Kolkata City. It is assumed that

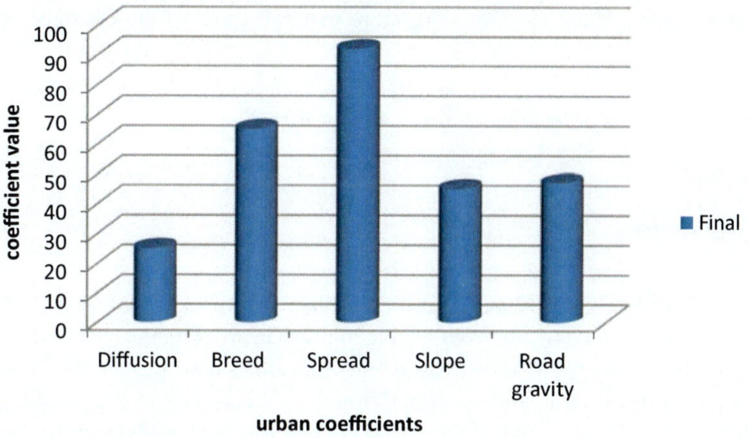

Fig. 6 Results of final calibration

urban expansion still persists according to the historical trends. It is also assumed that parks and water body were fully excluded from the future development. The results revealed that if the historical growth pattern will continue to unchanged in future, then the urban area of Kolkata City will reach to almost 70% of the study area by 2040. From the historical data, it is clearly seen that maximum urban development happened which proportion to reduction of natural ecosystem. The study also examined that in future 70–80% of non-urban area (barren land, cultivation land, and plants) will convert into the urban area due to increased population. Figure 7 displayed the urban growth simulated results of Kolkata City using SLEUTH model with 10 years interval from 2020 to 2040. From this Figure, it is visibly observed that in future more urban expansion will happen from 2020 to 2040 in the north-east and south-east directions of the Kolkata City which marked by red color in oval shape. From the outcomes, new spreading centers of urban growth were found in the north-eastern region at new town (Rajarhat division) location and south-eastern region at Garia-Sonarpur position of Kolkata City. In future, in these sectors will be more expanded as compared with

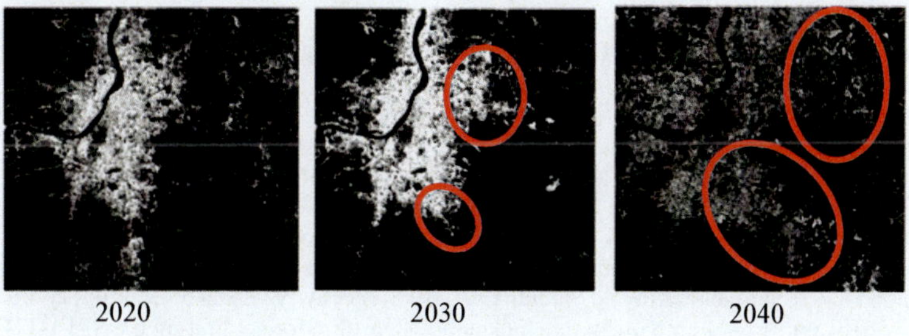

Fig. 7 SLEUTH simulation results of Kolkata City in 2020, 2030, and 2040

the others region. These locations are more expanded due to the communication, suitable slope, open land available, educational and the job oriented hub (IT sector hub) has been gradually increases day by day. Compact growth has been seen in the central Kolkata, along the Ganges River and its environs.

4 Conclusions

In this study, urban growth prediction of Kolkata City using CA-based SLEUTH model is presented. The result revealed that non-urban areas like barren land, plants, and agricultural area were transformed into the urban area due to the increased population. In future, if historical urban growth will continue in the coming years then 70% area will be occupied by urban area of the total study area in the year of 2040. From the historical urban scenario, it is noticeably observed that urban growth mainly led to the degradation of natural environmental resources. Study also revealed that in future more urban expansion may happen from 2020 to 2040 in the north-east and south-east positions of the Kolkata City. From this outcome, location-wise new urban growth center was found such as new town (Rajarhat division) in the north-east position and Garia-Sonarpur in the south-east position which spatially reflected in the map. The negative impacts of urbanization are increasing unplanned land transformation, improper development, poor air quality, and regional climate change in the city. Therefore, to proper planning and initiation is required for better and comfortable living standards of this city by 2040.

References

1. Clarke KC, Hoppen S, Gaydos L (1997) A self-modifying cellular automaton model of historical urbanization in the San Francisco Bay area. Environ Plan A 24:247–261
2. Khamchiangta D, Dhakal S (2019) Physical and non-physical factors driving urban heat island: case of Bangkok Metropolitan Administration, Thailand. J Environ Manage 248:109285
3. Ranagalage M, Wang R, Gunarathna MHJP, Dissanayake D, Murayama Y, Simwanda M (2019) Spatial forecasting of the landscape in rapidly urbanizing hill stations of south asia: a case study of Nuwara Eliya, Sri Lanka (1996–2037). Remote Sens 11:1743
4. Keeratikasikorn C, Bonafoni S (2018) Urban heat island analysis over the land use zoning plan of Bangkok by means of Landsat 8 imagery. Remote Sens 10(3)
5. Estoque R, Murayama Y (2017) Monitoring surface urban heat island formation in a tropical mountain city using Landsat data (1987–2015). ISPRS J Photogramm Remote Sens 133:18–29
6. Bhatta B (2009) Analysis of urban growth pattern using remote sensing and GIS: a case study of Kolkata, India. Int J Remote Sens 30(18):4733–4746
7. Kundu K, Halder P, Mandal JK (2020) Urban change detection analysis during 1978–2017 at Kolkata, India, using multi-temporal satellite data. J Indian Soc Remote Sens 48:1535–1554
8. Jafarnezhad J, Salmanmahiny A, Sakieh Y (2016) Subjectivity versus objectivity: comparative study between Brute Force method and Genetic Algorithm for calibrating the SLEUTH urban growth model. J Urban Plann Dev 142(3)
9. KantaKumar NL, Sawant NG, Kumar S (2011) Forecasting urban growth based on GIS, RS and SLEUTH model in Pune metropolitan area. Int J Geom Geosci 2(2):568–579

10. Rafiee R, Mahiny AS, Khorasani N, Darvishsefat AA, Danekar A (2009) Simulating urban growth in Mashad City, Iran through the SLEUTH model (UGM). Cities 26(1):19–26
11. Clarke KC (2008) Mapping and modelling land use change: an application of the SLEUTH model. In: Landscape analysis and visualisation, pp 353–366
12. Alsharif AAA, Pradhan B (2014) Urban sprawl analysis of Tripoli Metropolitan City (Libya) using remote sensing data and multivariate logistic regression model. J Indian Soc Remote Sens 42(1):149–163
13. Aburas MM, Ho YM, Ramli MF, Ash'aari ZH (2016) The simulation and prediction of spatio-temporal urban growth trends using cellular automata models: a review. Int J Appl Earth Obs Geoinf 52:380–389

Chapter 11
Suicide Ideation Detection in Online Social Networks: A Comparative Review

Sayani Chandra, Sangeeta Bhattacharya, Avali Banerjee(Ghosh), and Srabani Kundu

Abstract Online social network has turned out to have widespread existence on the Internet gradually. Social network services allow its users to stay connected globally, help the content makers to grow their business, etc. However, it also causes some possible risks to susceptible users of these media, for instance, the rapid increase of suicidal ideation in the online social networks. It has been found that many at-risk users use social media to express their feelings before taking more drastic step. Hence, timely identification and detection are considered to be the most efficient approach for suicidal ideation prevention and subsequently suicidal attempts. In this paper, a summarized view of different approaches such as machine learning or deep learning approaches, used to detect suicidal ideation through online social network data for automated detection, is presented. Also, the type of features used and the feature extraction methods for suicidal ideation detection are discussed in this paper. A comparative study of the different approaches to detect suicidal ideation is provided along with the shortcomings of the current works, and future research direction in this area is discussed in this paper.

Keywords Suicidal ideation detection · Online social network · Feature extraction · Machine learning · Deep learning · Natural language processing

1 Introduction

Approximately 8,00,000 people around the world die by suicide every year, which results in long-lasting effects on the families. It has been found that suicide is one of the important death causes among teenagers aged 15–29-year-olds [1]. Potential suicides happen due to the inability to deal with life stresses, such as financial or academic difficulties, stress due to relationship problems (such as break-ups or deaths of close ones) or stress due to harassment/ bullying. There is a strong connec-

S. Chandra (✉) · S. Bhattacharya (✉) · A. Banerjee(Ghosh) · S. Kundu
Guru Nanak Institute of Technology, Kolkata, India
e-mail: avali.banerjee@gnit.ac.in

J. K. Mandal et al. (eds.), *Proceedings of International Conference on Innovations in Software Architecture and Computational Systems*, Studies in Autonomic, Data-driven and Industrial Computing, https://doi.org/10.1007/978-981-16-4301-9_12

tion between suicide and mental disorders. Mental disorders including depression, schizophrenia, personality disorders, anxiety disorders, physical disorders such as chronic fatigue syndrome, and alcoholism are some of the major risk factors for suicide ideation. In addition, going through conflict, disaster, violence, abuse, or loss and a sense of isolation are also intensely associated with suicidal behaviour. Those who have previously attempted suicide are at a higher risk for potential future attempts [2].

Social media has slowly curved a position in our daily life in a profound way, by becoming an indispensable way of communication, especially during this pandemic situation. It has been found that people having suicide ideation prefer to communicate about their suicidal thoughts and plans over social media platforms than communicating face-to-face. Social media has many prospects to detect suicide ideation, by identifying the people at risk beforehand. Through social media, conversations and posts about suicide among at-risk people can be obtained. Researchers can analyse the same to provide some of the efficient and prospective approaches for suicide prevention using natural language processing and machine learning, [3]. Some social media platforms present suicide survivor stories in the form of animated avatars or text, which can also be used to identify keywords or emotions to detect suicidal tendency among people. In a nutshell, social media plays a major role in detecting and preventing suicides which has become one of the progressively significant fields of research [4].

This paper will provide an inclusive review of different suicidal ideation detection techniques. It intends to deliver a precise overview of the present research trends in this field and a viewpoint of the future research directions.

2 Literature Review

Suicide avoidance is possibly the best solution to decrease suicidal rates which can be implemented efficiently by identifying the suicidal thought-related text-based posts by users of any online social network. In [5], at first the sentiments of the tweets have been identified followed by the decision-making classification. Suicidal prediction has been done by lexicon and machine learning approaches. To recognize and discover the suicidal thoughts in user contents posted in any online social websites, a systematic content analysis, the language preferences and the topic representation from a data mining perspective have been accomplished in [6]. In [7], a combined model of long short-term memory (LSTM) and convolutional neural network (CNN) has been used to do classification for suicide ideation and communicate the knowledge of suicide ideation in the social network platform, Reddit. Suicidal thoughts have been detected using N-gram analysis. In [8], a feature stacking-based architecture known as Social Network Author Profiling-BiLSTM Attention NETwork (SNAP-BATNET) was proposed that used features such as author profiling, historical stylistic features, social network graph embeddings and tweet metadata from a large manually annotated dataset for detection of patterns of suicidal behaviour in online social media. To detect an impulsive behavioural changes in an individual Twitter user, a new application of a martingale framework has been used in [9], popular for identifying data stream setting changes. Also, Suicide Prevention Assistant (SPA) score for text analysis is proposed. In [10], the combination of sentiment analysis and linguistic features has been examined to learn a emotion model that represent suicidal thoughts and then automatically identified them based on supervised classification. In [11], it has been suggested to study text analysis practices to understand hidden users' emotions with the help of NLP techniques. The most advantageous linguistic features have been chosen in [12] and applied in depression identification to describe the post content in Reddit Social Media Forum. After that, they have analysed correlation significance using LIWC dictionary and the three feature types of linguistic dimensions, psychological actions and personal concerns, hidden topics using LDA method and word frequency, extracted from the text, using unigrams and bigrams where the vectors are based on TF-IDF scheme which are extracted from the text. In [13], suicidal tweets from Twitter's API were collected on which human encoding was used to decide the concern levels. Machine learning approaches have been used to automatically identify 'strongly concerning' tweets. Three deep learning methodologies have been used in [14] to detect suicide ideation from a labelled dataset generated from a lexicon of suicide-related words and phrases. STATENet: Suicidality assessment Time-Aware TEmporal Network, a novel framework that assesses the existence of suicidal intents on social media platforms by using a dual transformer-based architecture to learn the linguistic as well as the emotional signs found in tweets has been proposed in [15]. In [16], it has been proposed to use a classifier based on CNN model to reduce noise by filtering out unimportant tweets. After that, an RNN-based algorithm was used to identify the stress-related data from the suicide-related textual tweets. Two supervised learning classifiers, Naive Bayes and support vector machine, were used on the pre-processed dataset to classify into negative or neutral emotion state categories to detect depressive tweets in [17]. In [18], the tweets were classified into seven suicide-related categories which were used to test using machine learning classification algorithms to find out suicidal ideation in tweets. Weka as a data mining tool has been used in [19] to extract information for the classification of the tweets using several machine learning algorithms.

3 Methodology

Online social networks have proved to be an important platform to detect suicidal thoughts. People with suicidal considerations tend to use social media more nowadays to express their feelings. There are many popular social media platforms that people use to communicate their suicidal feelings with other users across the globe. Figure 1 shows the step-by-step process of suicide ideation detection in online social network.

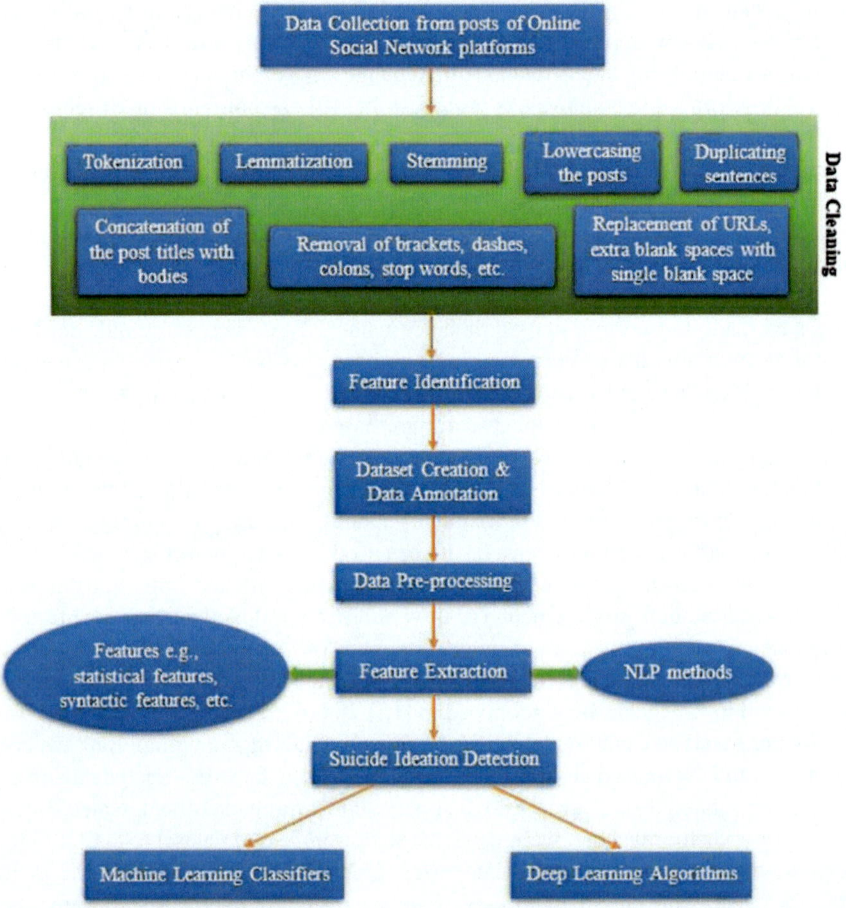

Fig. 1 A step-by-step diagram of suicide ideation detection in online social network

3.1 Data Collection from Online Social Network and Annotation

3.1.1 Data Collection from Online Social Network

In order to detect suicidal ideation, the first and foremost job is to collect the posts containing suicidal thoughts. Data from Twitter have been collected in [5, 8, 9, 13–19]. Dataset from Twitter has been collected based on annotation rule by using filtering technique utilizing suicidal words in [5]. Own dataset has been developed which is a two-phase process in [8]: (i) a dictionary of suicidal phrases was created and (ii) scrapping of tweets using the Twitter REST API using the lexicon. Twitter public streaming API has been used to collect tweets comprising key phrases related to suicide in [9, 14, 16, 18]. In order to study the development of the behavioural features, tweets have been identified where individuals showed a change in behaviour in [9]. The flow of tweets from such users is then used to examine the martingale method for detecting change point. Tweets have been collected using Twitter's public API in [13]. A collection of Twitter posts has been used as dataset built on a lexicon of 143 suicidal phrases in [15]. A dataset has been used that includes tweets collected using the Twitter API in [17] . The collection of tweets from Twitter has been done using Twitter4J in [19]. Data have been collected from Reddit social media platform in [7, 12]. Data have been collected from both Reddit and Twitter in [6]. Three datasets have been collected for analysis in [10]: one has original suicide notes, while the other two have been extracted from publicly made posts done in the Experience Project website. Data have been collected from Sina Weibo through Sina Microblog Open Platform API and their Java-based crawler in [11].

3.1.2 Data Annotation

After the data collection step has been done, the collected data have to be annotated. Data annotation has been done based on two classification schemes: (i) posts where suicidal intent present and (ii) posts containing no sensible confirmation that the possibility of suicide is present in [6–8, 14–16]. The collected data have been annotated using four classes of distresses from 0 to 3: (i) 0: no distress, (ii) 1: minimal distress, (iii) 2: moderate distress and (iv) 3: severe distress in [9]. Annotation of the collected data has been done using three levels of concerns: (i) strong concern, (ii) possible concern and (iii) safe to ignore in [13]. Annotation of the collected dataset has been done by classifying them into seven categories: (i) Class 1—possible suicidal intent evidence, (ii) Class 2—campaigning (such as petitions), (iii) Class 3—flippant reference to suicide, (iv) Class 4—support or information, (v) Class 5—condolence or memorial, (vi) Class 6—suicide reporting and (vii) Class 7—none of the above in [18]. Collected dataset has been annotated into two categories that consists of depression-indicative posts and standard posts in [12] .

3.2 *Preprocessing of Data*

The next step to data annotation is data pre-processing. Before moving to feature extraction, pre-processing must be performed. In pre-processing, input text is filtered to enhance the accurateness of the proposed system by removing excessive unnecessary features to process unprocessed data. To pre-process the data, some steps are to be followed. A sequence of filters is applied on collected Reddit platform posts to change unprocessed data into a proper format, compatible to the learning models in [7]. For this purpose, the natural language toolkit (NLTK) has been used. Firstly, it has been started with concatenation of the post titles with bodies and then duplicate sentences from the actual dataset have been removed. The next step is to do tokenization to filter out the data, followed by a conversion process to divide the Reddit posts into the separate tokens. Then, all the URL addresses, contractions and extra blank spaces have been replaced with a single blank space. After that, brackets, dashes, colons, stop words and all newline characters have been removed which can create unpredictable results if remains unnoticed. Thus, the posts have become lowercased and have been saved as separate files of texts. After that, lemmatization has been applied to confirm that the word endings are not unevenly fell down, which could create irrational word pieces such as stemming. That is why, they have been transformed into word lemmas which are related to the dictionary. In order to do pre-processing, a set of filters to decrease the noise has been employed in [8]. Firstly, a tweet tokenizer has been used to parse the tweet and replace every username mentions, hashtags and URLs with tokens. After that, stop words have been removed from the tokenized text which is then used as an input to WordNet Lemmatizer provided by NLTK. Finally, stemmed texts have been generated which are used as inputs to the feature extraction mechanisms using Lancaster Stemmer, provided by NLTK. NLP tools have also been used to pre-process the data in [12, 17]. Firstly, tokenization has been used to split the posts into separate tokens followed by removal of all the URLs, punctuations and stop words which can create erroneous results if not dealt with properly. Lastly, stemming has been applied to reduce the words to their root form and categorize alike words together. To improve the quality of the training data, the respective parts of speech have been assigned to the tokenized text by using POS Tagger in [17]. Pre-processing has been done by using a chain of filters to process the raw tweets in [14]. Firstly, non-English tweets have been removed followed by removal of URLs from the tweets. After that, user mentions in tweet bodies as well as retweets have been identified and eliminated. Then, all hashtags with length greater than 10 have been removed as a great volume of hashtags leads to redundant features. Three or more than three repetitive letters have been combined into a single letter followed by stop word removal and finally removal of tokens that are not a sequence of letters, - or ' which results in removal of numbers and terms that do not represent words. Data have been pre-processed by first removing email addresses, URLs and names in [15]. After that, texts have been converted to lowercase, punctuation and accents have been removed, whitespaces have been deleted, and stop words have been removed. Pre-processing of collected data has also been done by removing all the URLs and special keywords, e.g. 'hotline', 'suicide bomb', 'suicide attack', etc., from the tweets followed by removal of stop words in [16].

3.3 Feature Extraction

After pre-processing of the data, feature extraction is done on the cleaned data. Feature extraction is a concept that is concerned about transforming the raw data into inputs that a particular machine learning/deep learning algorithm requires. Features must signify the information of the data in a format that will best fit the needs of the algorithm that is going to be used to solve the problem. In this paper, we will summarize all the possible features and their extraction mechanisms relevant for suicide ideation detection.

3.3.1 Statistical Features

Statistical features have been extracted from texts in [5–7]. After segmentation and tokenization, statistical features are captured from the texts of the posts which includes the frequency of the tokens, characters, sentences and the length in the title and the body of the text.

3.3.2 Syntactic Features/Lexical Features

Parts of speech (POS) as part of syntactic features or lexical features have been extracted in [6, 8, 10, 11, 18, 19]. Conventional POS terms include nouns, participles, verbs, articles, adverbs, adjectives, pronouns and conjunctions. POS subcategories to present extra aspects about the grammatical characteristics of the posts have also been identified in [6]. Parsing and tagging have been done for each post, and the count of each category in the title as well as in the text body has also been taken. Natural language toolkit (NLTK) of Python to obtain POS tag information has been used in [10], whereas to extract POS tag statistics of each sentence, Stanford Parser Core NLP has been used in [11]. POS label has been assigned to each word in a tweet using Stanford POS Tagger in [18].

3.3.3 Linguistic Features

Linguistic features from the pre-processed data have been extracted in [6, 10, 12, 18]. Emotions, contingency and pestering words are generally contained in online users' posts. To extract these types of features, lexicons are broadly used. Linguistic Inquiry and Word Count (LIWC) is mostly used to study the emotional and linguistic features present in the data. To parse the data, LIWC encompasses a prevailing internally constructed dictionary to match the target words present in the posts. Apart from features based on word count, LIWC is also able to obtain features which are based on effective processes such as anxiety, +ve or −ve emotion and sadness; biological processes such as sexual, ingestion, body and health; social processes such as family, male, female and friend, cognitive processes such as cause, because, always and never; personal concerns such as job, cash, cook, kill and bury; and time orientations such as present, past and season. Apart from POS tag information, in terms of linguistic features, LIWC tool can also be used to extract note length, average length of sentence, contingency, denial, signs, e.g. +, &, adjectives, adverbs, verbs, etc.

3.3.4 Bag of Words or BoW

In BoW, the occurrence of each word in the pre-processed text is used as a feature for modelling a classifier [5, 7, 10, 14, 17]. The BoW model is a simplifying illustration which is used in NLP. In the working of this model, a text is represented as the bag of words present in it, calculating the frequency of each word in the document. The grammar and word order is disregarded, while multiplicity of words is kept. The frequency of words is used to generate a feature vector for training a given classifier. So, BoW is an algorithm that maintains a list of words along with the corresponding word counts per document.

3.3.5 Word Frequency Features

Word frequency features from the cleaned data have been extracted in [5–8, 13, 14, 18]. It measures the word occurrence frequency in a text. The significant words are selected, and words with low or no importance are eliminated. TF-IDF, short form of term frequency–inverse document frequency, is generally used to identify the features and measure the significance of various words found in suicidal posts. This is accomplished by multiplying the frequency of a word in a document and the inverse of the document frequency of the word across a set of documents.

3.3.6 Topic Features

Topic features from the dataset after pre-processing is done have been extracted in [6, 8, 12]. Topic modelling is a type of statistical modelling for uncovering the abstract 'topics' that could be found in a collection of documents. Topics associated with anxiety and depression that are hidden can be extracted from the selected documents using unsupervised text mining techniques. In case of LIWC, a set of pre-defined words is used. But here this is not the case. The group of unlabelled words is automatically produced. The words are selected based on a probability. As a result, each of the generated documents works with different topics that are linked with each other. Latent Dirichlet allocation (LDA) is an example of topic model and is used to categorize text in a document to a specific topic. It is a probabilistic generative model based on Dirichlet distributions that generates a topic for a document model and words for each topic model.

3.3.7 n-Gram Features

n-gram features from the pre-processed dataset have been extracted in [8–14, 18, 19]. An n-gram is a continuous sequence of n terms from a given text order. N-gram modelling, used to inspect the post features, is extensively used in text classification and NLP for detection of depression/suicide intent. It is very popularly used as the primary feature for post sentiment analysis. An n-gram can be of sizes 1, 2, 3, or any larger sizes, referred to as ‘unigram’, ‘bigram’, ‘trigram’, ‘four-gram’, ‘five-gram’ and so on, respectively, as the value of n. Python’s natural language toolkit (NLTK) to extract 2-g and 4-g has been used in [10], whereas Ansj, a Chinese word segmentation tool, has been used to extract unigram, bigram and trigram in [11].

3.3.8 Word Embedding Features

Word embedding features from the text have been extracted in [6, 14]. Word embedding is a set of feature learning techniques in natural language processing (NLP) where words or phrases from the dataset are mapped to vectors of real numbers. It is able to preserve the semantic information in texts. Word2vec is one of the popular techniques to produce word embeddings by generating a vector space from a large textual dataset. Since each unique word in the dataset is allocated an individual vector in the space, the vector is of many dimensions. GloVe or Global Vectors, developed by Stanford, is an unsupervised algorithm to generate word embeddings. It is used as an input to the next step in [8, 16]. In Glove, global word–word co-occurrence matrix is combined.

Apart from these prevalent features, some other features have also been extracted. NRC emotion lexicon which is a publicly available lexicon that encompasses frequently occurring words along with their affect category (anger, fear, anticipation, trust, surprise, sadness, joy or disgust) and two polarities (negative or positive) has been extracted in [8]. The count of hashtags, mentions, URLs and emojis along with the retweet count and favourite count of every tweet have also been extracted as part of tweet metadata features. node2vec algorithm has been used for converting nodes in a weighted or unweighted graph into feature representations which has been applied on all the social graphs. To identify online behaviours that may reveal the mental state of a Twitter user, two groups of behavioural features user-centric and post-centric features have been established in [9]. User-centric features characterize the behaviour of the user in the Twitter community, e.g. friends, followers, volume, replies, retweets, links, questions, while post-centric features are characteristics that are extracted from the properties of a tweet, e.g. time and text score. It has been found that temporal features do matter a lot in [11]. Temporal features may involve the time posts, the type of posts, etc. Posting time refers to the time when the post was made, whereas posting type indicates original posts or forwarded posts. Another feature that has been considered by them is posting frequency which indicates the frequency of posts as time varies. 768-dimensional encoding obtained from SentenceBERT and Plutchik transformer-based encodings has been used in [15]. Features signifying idiosyncratic language communicated in short, informal text such as social media posts within a restricted number of characters have been used in [18].

3.4 *Classification of Texts*

After the features are extracted, they are fed as input to the classification algorithms in order to identify whether suicidal intents are present or not in those features. For this purpose, several machine learning classification algorithms as well as deep learning algorithms are used. Four metrics, namely accuracy, precision, recall and F1 score, are generally used for the classification task.

3.4.1 Machine Learning Algorithms

In [5], Naive Bayes classifier (NB), random forest (RF), eXtreme Gradient Boosting (XGBoost) and logistic regression (LR) have been used to predict the suicidal ideation. NB and LR have shown alike results in precision and recall. When different features in NB are applied, a reduced amount of variation in accuracy can be seen. XGBoost and RF classifiers have given alike performances with precision and F1 value, whereas if all the features are combined then RF method has been seen to pull off improved accuracy than XGBoost. Therefore, by using classifier algorithms on the datasets, the suggested model has accomplished 0.99% accuracy, 0.96% precision, 0.91% recall and 0.98% F1 value. In [9], concentration has been given to detect distress-related and suicidal contents and established two approaches to measure a tweet: an approach based on NLP and a text classifier based on machine learning. The testing has been done with eight different classification algorithms, namely multinomial NB, C4.5 decision tree (J48), multinomial LR, nearest neighbour classifier (IB1), rule induction (Jrip), sequential minimal optimization (SMO) with a poly kernel, RF and SMO with a Pearson VII universal kernel (PUK) function. It has been noticed that SMO with a PUK function has performed the best in terms of precision as it produces 66.4% value with n-grams, symptoms, pronouns and swear component features. To detect changes in behaviour, they have used martingale framework to detect change points within a Twitter user's activity stream. The NLP-based approach that they have opted acts as an input into the martingale framework which can successfully distinguish whether tweets are exhibiting distress-related content or not. In [10], the WEKA toolkit has been used for their supervised learning researches. Logistic tree regressor or logistic model tree (LMT), NB classifier, J48 decision tree and a simple majority baseline classifier (Zero-R) have been used by them. It has been observed that LMT regressor has given the best result with 86.61% of overall accuracy whenever sentiment and linguistic features are used together. In [11], SVM classifier from LibSVM package for C and Java has been chosen to be used along with several classifiers from Weka which are NB, LR, J48, RF and SMO. It has been observed that SVM classifier has achieved 68.3% f-measure value, 60.3% recall value and over 94% accuracy value. The accuracy goes over 94% if all the posts are considered. In [12], five machine learning algorithms, LR, SVM, RF, adaptive boosting (AdaBoost) and multilayer perceptron (MLP) models, have been used for the classification task. It has been observed that SVM has given best result when used with bigrams which results in 80% accuracy and 0.79 F1 score. LIWC feature with RF model gives 78% accuracy and 0.84 F1 score, and finally, LDA feature with LR model causes 77% accuracy and 0.83 F1 score. It has been found that LIWC with RF outperforms SVM and MLP in terms of accuracy and F1 score. The combined feature set LIWC+LDA+bigram with MLP has accuracy of 91% and F1 score of 0.93 outperforming all the other classifiers with all the other possible combination of features. In [13], two machine learning classifiers SVM and LR have been used. It has been seen that the best performance was given by SVM with TF-IDF no-filter. It has been observed that when both the training and testing tweet sets were integrated, a performance gain in accuracy of 76% was obtained. In [17], two classifiers, multinomial NB and SVM, have been used for the classification task. It has been observed that multinomial NB performed best while SVM performance was low with F1 score of 83.29 and 79.73, respectively. The same tendency can be seen in terms of precision and recall metrics also where multinomial NB outperforms SVM. It has also been observed that accuracy of the multinomial NB is 83 while 79% in case of SVM. In [18], an ensemble approach known as rotation forest (RF) has been proposed and compared with three other baseline models, NB, J48 DT and SVM. It has been found that the rotation forest algorithm has achieved an overall F1 score of 0.728 for 7 classes which includes suicidal thought and 0.69 only for the class suicidal ideation. In [19], five well-known classification algorithms available in WEKA, IB1, J48, Classification and Regression Trees (CART), SMO and NB have been used. While evaluating the suspected tweets with risk of suicide, SMO has outperformed all the other classifiers, whereas in case of evaluating the suspected tweets without risk of suicide, NB has outperformed all other models in terms of recall and F1 score, but in terms of precision metric J48 decision tree has achieved the best performance value, 75.4%. Table 1 summarizes the above methodologies and their drawbacks.

3.4.2 Deep Learning Algorithms

In [8], a novel feature stacking-based architecture, SNAP-BATNET (Social Network Author Profiling-BiLSTM Attention NETwork) has been proposed that uses author profiling, historical stylistic features, social network graph embeddings and tweet metadata features. It has been observed that among all the other deep learning/machine learning models, SNAP-BATNET, when linked with all other feature sets using feature stacking, achieved the best outperforming the current state-of-the-art approaches. After collecting and filtering suicide-related tweets using keywords, CNN has been used to remove more noise by further filtering out inappropriate tweets followed by the use of RNN to extract the stress or mentions from the suicide-related tweets in [16]. The impact of transfer learning-based approaches has also been inspected. It has been found that CNN model performed better than RNN-based framework from the F-measure of 83% and 53.25%, respectively. It has been observed that the same performance can be achieved by transfer learning, thereby reducing the cost of annotating the tweets.

Table 1 A comparative analysis of different suicide ideation detection methodologies in various online social network

References	Features extracted	Methodology used for suicidal ideation detection	Social network involved	Drawback(s)
[5]	Statistical features, BoW, TF-IDF, distributed features	NB, RF, XGBoost, LR	Twitter	–Temporal (historical) data not used
[6]	Statistical features, POS, LIWC, TF-IDF, Word2vec, LDA	SVM, RF, GBDT, XGBoost, MLFFNN, LSTM	Reddit and Twitter	–Temporal (historical) data not used
[7]	Statistical features, TF-IDF, BoW	LSTM-CNN	Reddit	–Correlation between suicidal thought and environmental aspects is absent –Supervised algorithms may be costly in terms of time and possible annotator inaccuracy
[8]	TF-IDF, POS, GloVe embeddings, NRC emotion, LDA, Tweet metadata features, GloVe embeddings, NRC sentiment scores and POS, node2vec	SNAP-BATNET	Twitter	–Multi-modalities in the data in the form of images, videos and hyperlinks were absent
[9]	Behavioural features: user-centric and post-centric features, n-grams	Martingale framework for emotion change detection	Twitter	–Dataset consists of only two Twitter users' data –Coarse-grained sentiment classes such as anger, sadness and fear, not explored
[10]	BoW, POS, n-grams, LIWC	LMT, J48, NB, Zero-R	Experience project	–Not tested on presently used popular social networking sites

(continued)

Table 1 (continued)

References	Features extracted	Methodology used for suicidal ideation detection	Social network involved	Drawback(s)
[11]	n-grams, POS, temporal features (posting time and post's type), posting frequency	SVM, NB, LR, J48, RF, SMO	Weibo	–Most likely social network correspondence to identify suicide individuals and groups not explored –Dataset not popularly available
[12]	LIWC, LDA, n-grams	LR, SVM, AdaBoost, RF, MLP	Reddit	–Association between the users' character and their depression-related activities reflected in social media is not examined
[13]	TF-IDF, unigrams	SVM, LR	Twitter	–Twitter data with sample features, such as age and gender, are not used –Uncommon or curse words which represent high risk of suicide are not included
[14]	Word2vec, character n-grams, TF-IDF, BoW	RNN, LSTM, C-LSTM	Twitter	- Could not detect suicide ideation in tweets containing subtle references, uncertainty, unfamiliarity - Nature-inspired heuristics need to be explored for efficient feature selection

(continued)

Table 1 (continued)

References	Features extracted	Methodology used for suicidal ideation detection	Social network involved	Drawback(s)
[15]	768-dimensional encoding obtained from SentenceBERT, Plutchik transformer-based encodings	STATENet	Twitter	–Quantification of impact of varying degrees of granularity of learning emotional features from tweets is missing
[16]	GloVe Twitter embedding	CNN, RNN	Twitter	–Historical tweets to access the user-level information are missing
[17]	BoW	Multinomial NB, SVM	Twitter	- High false positives
[18]	TF-IDF, n-grams, proposed feature set 1, feature set 2, feature set 3, data-driven features	Rotation forest (RF)	Twitter	–Concepts and labels representing broader semantic domains are mostly generated by a 'confusion' and 'misrepresentation' of words –Inclusion of sentiment analysis. Scores are barely included within the primary features of each class
[19]	Term presence, term frequency, negation, n-grams, POS	IB1, J48, CART, SMO, NB	Twitter	–Multilingual WordNet for tweets can be tested

3.4.3 Combination of Machine Learning and Deep Learning

In [6], support vector machine (SVM), gradient boosting decision tree (GBDT), multilayer feedforward neural network (MLFFNN), long short-term memory (LSTM) models along with RF and XGBoost models have been used. XGBoost has outperformed all the other models in terms of the above-mentioned metrics when all groups of features are fed as input. It has been observed that RF, GBDT, XGBoost and MLFFNN with appropriate features generated improved accuracy and F1 values compared to LSTM. LSTM with word embedding feature has given better precision result compared to the others. In [7], a combination of LSTM and convolutional neural network (CNN) has been used as a proposed model along with six other baseline models SVM, NB, RF, XGBoost, LSTM and CNN. It has been observed that XGBoost has scored better than other established text classification methods when used with combined as well as single features, except the statistical features. It has also been observed that the combined model of LSTM-CNN with word embedding feature has outperformed other traditional algorithms of machine learning as well as deep learning algorithms as it reaches accuracy of 93.8% and F1 value of 92.8%. In [14], deep learning models recurrent neural network (RNN), CNN and C-LSTM along with two other baseline models SVM and LR with character n-grams, TF-IDF and bag of words features have been used. It has been seen that C-LSTM has performed considerably better than the two baseline methods as well as vanilla LSTM and RNN. RNNs were as good as both of the TF-IDF and bag of words features combined with multinomial LR and SVMs. Among the baselines, the TF-IDF features along with multinomial LR have performed better than others. The proposed model STATENet was based on time-aware transformer method [15]. Testing of their proposed model has been done against seven baseline models, RF, C-LSTM, suicide detection model (SDM), contextual CNN, exponential decay, surprise and episodic modelling and DualContextBert. It has been observed that STATENet has notably outperformed likely baseline models. STATENet and other contextual models have shown better performance than the non-contextual RF and C-LSTM models. STATENet and sequential models have outperformed the contextual CNN. STATENet has also considerably outperformed the sequential models, SDM and DualContextBert. It has also been observed that STATENet performs better, in terms of all metrics, especially recall for the suicidal intent present class.

4 Open Research Problems

Although several researches have been carried out on suicide ideation detection in online social networks, there are still some areas where improvement can be done.

1. Temporal data or historical data are very effective in identification of suicidal ideation. However, most of the works [5, 6, 16] have not explored the same; instead, only single post data without any context are used. Also, the relationship between suicidal thought and environment needs to be studied for improving the performance of the detection methods.
2. Automatic labelling of the data is preferred to avoid annotation bias produced by manual categorization with some predetermined annotation rules.
3. Multi-modalities in the data in the form of images, videos and hyperlinks also need to be analysed along with textual data to detect suicidal ideation.
4. A dataset needs to be created using different social network data consisting of features extracted from various types of posts such as text, image and video to

provide the researchers a common platform to test the efficiency of the proposed model.

5. Instead of supervised learning classification technique, a layer expert-based suggestion can be applied to the used model to decrease the number of false positives, thereby increasing the precision of suicidal ideation detection.

References

1. World Health Organization (2018) National suicide prevention strategies: progress, examples and indicators. World Health Organization, Geneva, Switzerland
2. W. h. Organization (2014) Preventing suicide: a global imperative, website, 2014, http://www.who.int/mentalhealth/suicide-prevention/en/
3. Parrott S, Britt BC, Hayes JL, Albright DL (2020) Social media and suicide: a validation of terms to help identify suicide-related social media posts. J Evid Based Soc Work 17(5):624–634
4. Luxton DD, June JD, Fairall JM (2012) Social media and suicide: a public health perspective. Am J Public Health 102(S2):S195–S200
5. Rajesh Kumar E, Rama Rao K, Nayak SR, Chandra R (2020) Suicidal ideation prediction in twitter data using machine learning techniques. J Interdisc Math 23(1):117–125
6. Ji S, Yu CP, Fung S-F, Pan S, Long G (2018) Supervised learning for suicidal ideation detection in online user content. Complexity
7. Tadesse MM, Lin H, Xu B, Yang L (2020) Detection of suicide ideation in social media forums using deep learning. Algorithms 13(1):7
8. Mishra R, Sinha PP, Sawhney R, Mahata D, Mathur P, Shah RR (2019) Snap-batnet: cascading author profiling and social network graphs for suicide ideation detection on social media. In: Proceedings of the 2019 conference of the North American Chapter of the Association for computational linguistics: student research workshop, pp 147–156
9. Vioules MJ, Moulahi B, Azé J, Bringay S (2018) Detection of suicide-related posts in twitter data streams. IBM J Res Dev 62(1):7–1
10. Schoene AM, Dethlefs N (2016) Automatic identification of suicide notes from linguistic and sentiment features. In: Proceedings of the 10th SIGHUM workshop on language technology for cultural heritage, social sciences, and humanities, pp 128–133
11. Huang X, Zhang L, Chiu D, Liu T, Li X, Zhu T (2014) Detecting suicidal ideation in Chinese microblogs with psychological lexicons. In: IEEE 11th international conference on ubiquitous intelligence and computing and 2014 IEEE 11th international conference on autonomic and trusted computing and 2014 IEEE 14th international conference on scalable computing and communications and its associated workshops. IEEE, pp 844–849
12. Tadesse MM, Lin H, Xu B, Yang L (2019) Detection of depression-related posts in reddit social media forum. IEEE Access 7:44883–44893
13. O'dea B, Wan S, Batterham PJ, Calear AL, Paris C, Christensen H (2015) Detecting suicidality on Twitter. Internet Interv 2(2):183–188
14. Sawhney R, Manchanda P, Mathur P, Shah R, Singh R (2018) Exploring and learning suicidal ideation connotations on social media with deep learning. In: Proceedings of the 9th workshop on computational approaches to subjectivity, sentiment and social media analysis, pp 167–175
15. Sawhney R, Joshi H, Gandhi S, Shah R (2020) A time-aware transformer based model for suicide ideation detection on social media. In: Proceedings of the 2020 conference on empirical methods in natural language processing (EMNLP), pp 7685–7697
16. Du J, Zhang Y, Luo J, Jia Y, Wei Q, Tao C, Xu H (2018) Extracting psychiatric stressors for suicide from social media using deep learning. BMC Med Inf Decis Making 18(2):43
17. Deshpande M, Rao V (2017) Depression detection using emotion artificial intelligence. In: 2017 international conference on intelligent sustainable systems (ICISS), IEEE, pp 858–862

18. Burnap P, Colombo W, Scourfield J (2015) Machine classification and analysis of suicide-related communication on Twitter. In: Proceedings of the 26th ACM conference on hypertext & social media, pp 75–84
19. Birjali M, Beni-Hssane A, Erritali M (2017) Machine learning and semantic sentiment analysis based algorithms for suicide sentiment prediction in social networks. Procedia Comput Sci 113:65–72

Chapter 12
An Improved K-Means Algorithm for Effective Medical Image Segmentation

Amlan Dutta, Abhijit Pal, Mriganka Bhadra, Md Akram Khan, and Rupak Chakraborty

Abstract Clustering-based image segmentation got wide attention for decades. Among various existing clustering techniques, K-means algorithm gained popularity for its better outcome. But the drawback of this algorithm can be found, when it is applied to noisy medical images. So, modification of the standard K-means algorithm is highly desired. This paper proposes an improved version of K-means algorithm called as (**IKM**) to get more effective and efficient outcomes. The efficiency of the algorithm depends on the speed of forming the clusters. So, in the proposed approach, new idea has been applied to find the minimum distance to generate the clusters. The proposed IKM algorithm has been applied to the set of noisy medical images, and the segmented outcomes have been evaluated by the standard quality measurement metrics, namely Peak-Signal-to-Noise-Ratio (PSNR) and structural similarity index measurement (SSIM). The outcomes have also been compared with the Watershed algorithm for showing the betterment of the proposed approach.

Keywords Segmentation · K-means · Watershed · PSNR · SSIM

1 Introduction

Every image is a collection of small square blocks known as pixels which are stored in an 2-D array. Images are basically of three types on the basis of Bit depth: Black and White image, Grayscale image, Color image. In this paper, we will be working on Grayscale image. A Grayscale image contains only different sets of Gray. Grayscale image stores the intensity as an integer of 8-bit, which gives 256 different available shades of Gray from Black to White. Image segmentation is a process of classifying the pixels of an image. It can be achieved by splitting an image into some discrete regions with higher similarity in each region and higher contrast between regions. It analyzes what is inside an image, which is used in various fields like Image

A. Dutta · A. Pal · M. Bhadra · M. A. Khan · R. Chakraborty (✉)
Department of Computer Science and Engineering, Guru Nanak Institute of Technology, Kolkata, India
e-mail: rupak.chakraborty@gnit.ac.in

J. K. Mandal et al. (eds.), *Proceedings of International Conference on Innovations in Software Architecture and Computational Systems*, Studies in Autonomic, Data-driven and Industrial Computing, https://doi.org/10.1007/978-981-16-4301-9_13

Processing, Healthcare, Pattern Recognition, Feature Extraction, etc. Image segmentation has different approaches like threshold based, edge based, cluster based, neural network based. In these different techniques, clustering is one of the most efficient methods, which has different types: K-means clustering, Mountain clustering, fuzzy C-Means clustering and Subtractive clustering, etc.

K-means clustering is one of the mostly used clustering methods. It is computationally faster than other clustering methods available. This simple clustering method is able to work with large number of variables. It is needed to initialize the proper number of cluster k2 because it produces different results for different number of clusters. Selecting the proper initial centroid is also a prime task because different values of initial centroid would result in different cluster.

Section 2 represents the related work of the papers done in the field of K-means. Section 3 formulates the problem of K-means clustering. Section 4 is the proposed IKM algorithm. The results of the proposed algorithm along with other considered optimized methods are presented in Sect. 5. Section 6 represents the performance analysis of the improved K-means algorithm. Finally, Sect. 7 draws a conclusion of the paper with its limitation and future scope.

2 Literature Review

K-means clustering algorithm is one of the most commonly used algorithm for image segmentation. Many works already exist in this field done by researchers. Some works are discussed here:

Different no. of clusters are performed on color images using K-means [1–3]. An image is converted from RGB to HSV color model by Jumb et al. [4]. Taking the V value, multi-thresholding is applied, since it is directly proportional to intensity of the image. Different clustering are applied: K-means and Subtractive, generating initial center for segmentation of image and removing unwanted region [5, 6]. To decrease the light effect on segmentation of image, transformation of color space is done to convert into LAB color space[7].

Image segmentation is compared by fuzzy C-Means clustering and reviewed [8]. By the use of K-means and fuzzy K-means huge amount of time and data can be saved [9, 10]. Controlling parameters for evolution set lead to fuzzy clustering [11–13]. By using K-means clustering with decision tree, many Business Customers have been segmented [14, 15]. In medical field, a well-known device called MRI is used in this principle to segment images from one another [16–19]. In some cases, CNN is used for segmentation for trained data [20].

Though K-means clustering has some drawbacks such as predicting the K-value becomes difficult. It did not work well with global cluster and different size or density. Different final clusters are resultant of different initial partitions. But, as an improvement, K-means clustering algorithm has been significantly improved by using better initialization technique, and by iterating, i.e., repeating the algorithm [21].

Now, by inspiring from the work done previously, we have proposed an Improved K-means algorithm (IKM), which is more efficient. We have also compared our proposed algorithm, with the latest modified K-means such as Adaptive K-means (AKM) [22] and K-means initialization Technique [21]. We have put together all the data and checked the time comparisons between them and presented them in different forms of graphs.

3 Problem Formulations

K-means clustering method is one of the unsupervised learning methods which is applied to solve low-level image segmentation tasks. K-means clustering is similar to nearest neighbor techniques (collaborative filtering and memory-based reasoning). It is a very important task to choose the initial cluster center's in case of K-means clustering because this prevents the algorithm from producing incorrect decisions. The procedure of K-means clustering is given below.

3.1 Procedure

K-means clustering technique is an efficient clustering technique. It is used to separate initial data into groups based on initial centroids chosen by the cluster. K-means algorithm is also called distance-based algorithm. According to this algorithm, k data values are being chosen as initial clusters and then it finds the distance between each cluster center and each data value then it assigns the value to the nearest cluster, update the average of every cluster and keep repeating this process until the criterion does not match. Figure 1 shows the process of basic k-means.

K-Means Algorithm: In case of K-means clustering algorithm, center of each cluster is represented by mean value of its objects in the cluster.

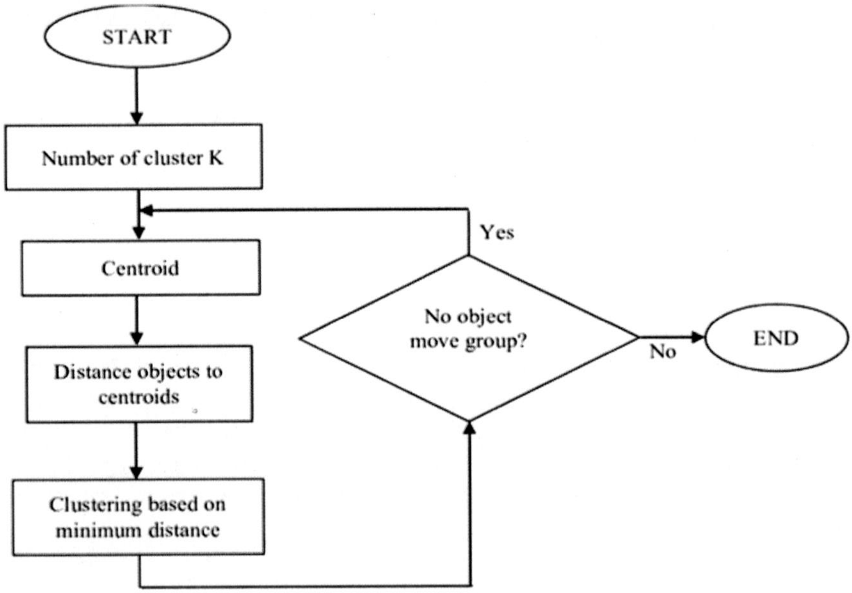

Fig. 1 Process of K-means

Input :

K represents total number of clusters and D represents data set having n objects.

Output : A set of k clusters.

Method : Choose initial values for each u (from $u = 1$ to k);

```
repeat {
  Assign input data for each point to the closest u,

 for each u calculate the new mean value;

  if all u values are unchanged
  {
    break;
  }
}
```

$$J = \sum_{j=1}^{k} \sum_{i=1}^{n} x_i^{(j)} - c_j^2$$

Here, $x_i^{(j)} - c_j^2$ is the calculated selected distance among the point x_i and the center of the cluster c_j. The cluster center is an indicator of the distance of cluster center from their n data points.

We have to specify the number of clusters as an input to the algorithm which is one of the main disadvantages of K-means. The algorithm is not able to find the appropriate number of clusters, and it depends upon the user to identify this in advance.

4 Proposed IKM Algorithm

There are a lot of research works that have been done previously based on the K-means clustering algorithm. Those all are done based on the improvement of the clustered result. Based on those works, we get influenced, and here, we decided to improve the algorithm a bit so that it can work more fast.

There are some basic works of K-means clustering algorithm what we have done also here they are find the initial centroid. In our work, we have tried to minimize the calculation of the K-means algorithm so that the algorithm will work fast. After merge all the improvements, we get the final faster and modified K-means algorithm.

Input:

$S = \{n_1, n_2, n_3, \ldots n\}$ // Set of n number of data points (Our case it is the matrix form of the image).

Output:

A set of clusters.

Steps:

1. First calculate $R \cong \sqrt{\frac{n}{2}}$.
2. Let $x = 1$.
3. To select the value of 'f' follow the following condition

if the value of f is not given then $f = 1$ and $f > 0$.

4. Now detect nearest pair of data points from the set S. Make a new set N_x with those points.
5. Now select all the point from S which are nearest to N_x and shift them to N_x from S.
6. Keep repeating step 5 till the total number of element of N_x set goes to $\frac{n}{R}$ Xf
7. At the point of time N_x are full and the total count of elements of $S \neq 0$, then increment the value of x by 1 ($x = x + 1$). Keep repeating step 4.
8. Calculate center of gravity of each set N_x. These are the initial centroids C_j and values of N_x will be the elements of R_m.
9. Now detect the nearest centroid for each data points and allocate each data points cluster. Compute the new centroids for each R_m.
10. Compute D_1 which is the **largest distanced** data point from each centroid of each Cluster.

11. Obtain the value points in the range of D_1 X 4/9 to D_1.
12. Now detect the closest centroid for all those points and assign these points to their nearest centroid cluster.
13. Detect the new centroid values for each R_m.
14. If there is any modification in any R_m, then keep repeating the step 10 or stop the algorithm.

5 Experimental Results

5.1 Experimental Setup

We have designed a modified version of the standard K-means algorithm. To experiment this and also for development purposes, we have used Python version 3.8 Jupyter Notebook version 6.2.0, OpenCV version 4.5.1.48 for reading the image, matplotlib library version 3.3.4 for plotting the segmented images. We have used the **Grayscale** images for our experiments. To obtain grayscale image of the standard medical image, we have used the rgb2gray method of scikit-learn library then we have used our algorithm. Used Datasets:

(1) https://wiki.cancerimagingarchive.net/pages/viewpage.action?pageId=70229053
Size: 26 GB.
(2) https://wiki.cancerimagingarchive.net/display/Public/CT+COLONOGRAPHY#e88604ec5c654f60a897fa77906f88a6
Size: 462.6 GB.
(3) https://www.kaggle.com/kmader/siim-medical-images
Size: 2 GB.

5.2 Compared Images

Based on the modified K-means algorithm, some experimental results have been listed below.

6 Performance Analysis

The PSNR or Peak-Signal-to-Noise Ratio is the ratio between maximum possible power of an image and the power of corrupting noise. PSNR has a good affection on representing an image. To estimate, the PSNR of an image is it necessary to compare the image with an ideal clean image.

PSNR can be defined as follows:

$$P_{\mathrm{PSNR}} = 10\log_{10}\left(\frac{(L-1)^2}{\mathrm{MSE}}\right) = 20\log_{10}\left(\frac{(L-1)}{\mathrm{RMSE}}\right)$$

Here, L represents the value of highest possible intensity levels (lowest intensity level assumed as 0 for an image).

The **SSIM** or **structural similarity index measure** technique is used to predict the perceived quality of digital TV and pictures. We are able to measure the similarity between two images with SSIM technique. The SSIM index is a full reference metric, i.e., the measurement of prediction of image quality is based on an initial uncompressed image.

SSIM can be defined as follows:

$$S_{\mathrm{SSIM}}(x, y) = \frac{(2\mu_x\mu_y + c_1)(2\sigma_{xy} + c_2)}{(\mu_x^2 + \mu_y^2 + c_1)(\sigma_x^2 + \sigma_y^2 + c_2)}$$

Here,

The average value of x is represented by μ_x.
The average value of y is represented by μ_y.
The variance of x is represented by σ_x^2.
The variance of y is represented by σ_y^2.
The covariance of x and y is represented by σ_{xy}.
Here, $c1$ and $c2$ are used as variables to stabilize the division which has weak denominator.
Here, $c_1 = (k_1L)^2$ and $c_2 = (k_2L)^2(k_2L)^2$.
The dynamic values of pixel values are defined by L.
The default value of k_1 and k_2 is 0.01 and 0.03, respectively.

Here, MATLAB R2018 workstation is used for simulation in Computer with configuration of core i5 3.2 GHz processor and 8 GB of RAM. The setup of AKM [22] and KIT [21] has been chosen by following the guidelines given in the above literature and noted. Table 1 represents the comparison between IKM, AKM and KIT using PSNR and SSIM index values of level 5, 7 and 11. Table 2 represents the time comparison (t) between them. Figure 1 represents the segmented images for different optimizers using K-means. In Fig. 2, we have found out the time graph of our Improved K-means and AKM, KIT represented by Red, Green and Blue bars. Also observed is that the optimized time is taken for IKM for level 7 and 11 values, given by the Red bar. Now, comparisons are made between IKM, AKM [22] and KIT [21] using PSNR and SSIM index values of level 7 and 11 box plots. The box plots represent various box size for different values of level 7 and 11. Evaluation of measurements is carried out in the form of comparison time (t), whereas the quality of segmented grayscale images is measured by the popular metrics, namely PSNR and SSIM.

Here, in Table 1, PSNR and SSIM are the parameters to detect whether an algorithm is good or not and more the values of PSNR and SSIM, better is the result. The

Table 1 Comparison of PSNR and SSIM index values between IKM, AKM and KIT

I_m		5-level			7-level			11-level		
		IKM	AKM	KIT	IKM	AKM	KIT	IKM	AKM	KIT
1	PSNR	30.630	27.330	28.63	31.42	29.03	30.30	33.56	30.7	31.825
	SSIM	0.620	0.518	0.568	0.73	0.68	0.712	0.859	0.792	0.832
2	PSNR	30.650	27.191	28.736	31.756	29.131	30.455	33.629	31.428	31.955
	SSIM	0.624	0.518	0.586	0.756	0.712	0.755	0.868	0.815	0.882
3	PSNR	30.748	28.157	28.789	31.955	29.325	30.546	33.739	31.747	31.974
	SSIM	0.629	0.529	0.596	0.780	0.761	0.789	0.877	0.817	0.896
4	PSNR	30.530	27.259	28.610	31.425	28.824	30.121	33.613	31.440	31.943
	SSIM	0.608	0.504	0.515	0.721	0.735	0.725	0.866	0.805	0.881

Table 2 Time comparison (t) between IKM, AKM and KIT

I_m		5-leve			7-level			11-level		
		IKM	AKM	KIT	IKM	AKM	KIT	IKM	AKM	KIT
1	t	2.618	5.329	3.218	4.017	8.306	6.198	5.698	9.998	7.528
2	t	2.692	5.357	3.275	4.047	8.332	6.223	5.733	10.025	7.556
3	t	2.710	5.387	3.315	4.098	8.384	6.256	5.762	10.048	7.581
4	t	2.690	5.350	3.286	4.050	8.358	6.212	5.722	10.032	7.530

comparison shown in Table 1 shows that PSNR and SSIM of Improved K-means (IKM) algorithm are higher than both Adaptive K-means (AKM) algorithm and K-means initializing Technique (KIT) in 5-level, 7-level and 11-level in all 4 cases in the table.

Table 2 shows time comparison(time complexity) among Improved K-means (IKM) algorithm, Adaptive K-means (AKM) algorithm and K-means initializing Technique (KIT) and in all 4 cases of 5-level, 7-level and 11-level the time taken by Improved K-means (IKM) algorithm is lesser than both Adaptive K-means (AKM) algorithm and K-means initializing Technique (KIT).

Bar Graph

Here, three bar graphs show relationship between comparison time in seconds and different images in all 4 cases with respect to Improved K-means (IKM) algorithm is lesser, Adaptive K-means (AKM) algorithm and K-means initializing Technique (KIT) at Lv = 7 time and Lv = 11 time, and in every cases, the time complexity of Improved K-means (IKM) algorithm is lesser than both Adaptive K-means (AKM) algorithm and K-means initializing Technique (KIT) (Fig. 3).

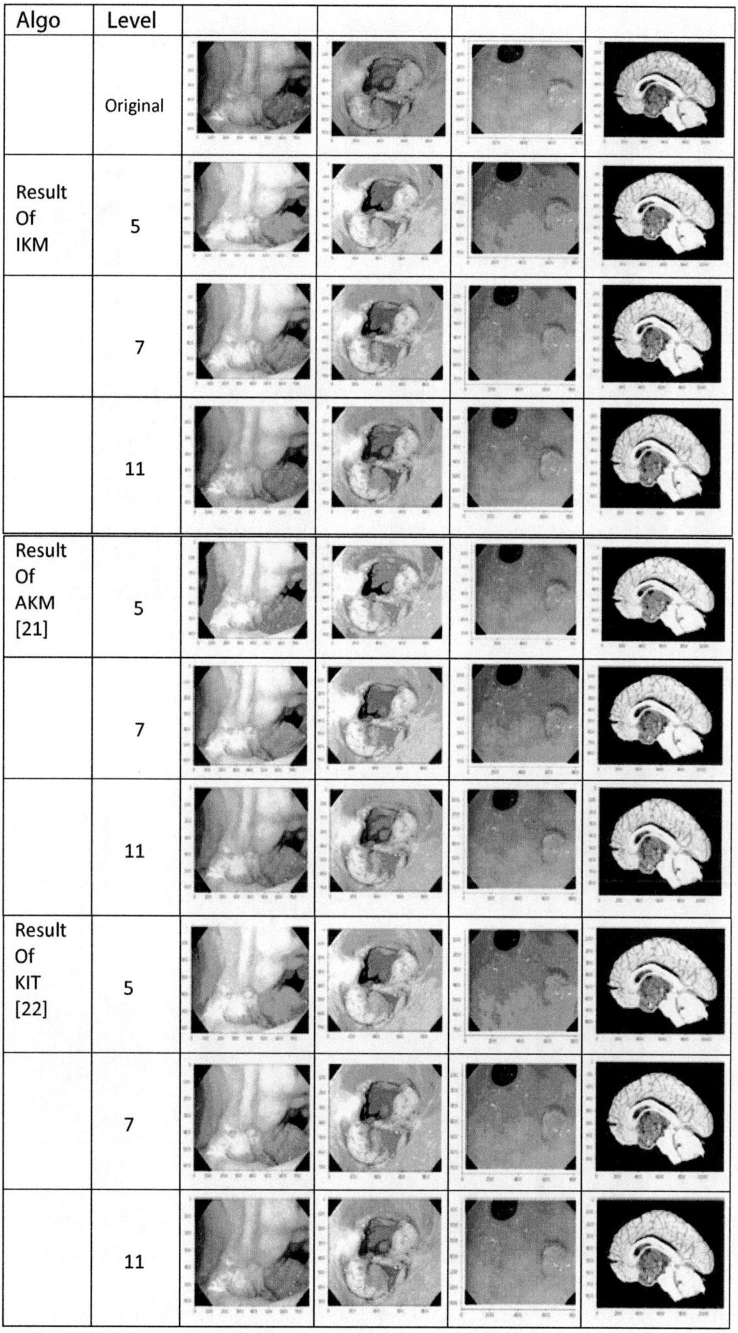

Fig. 2 Segmented images with different optimizers using K-means

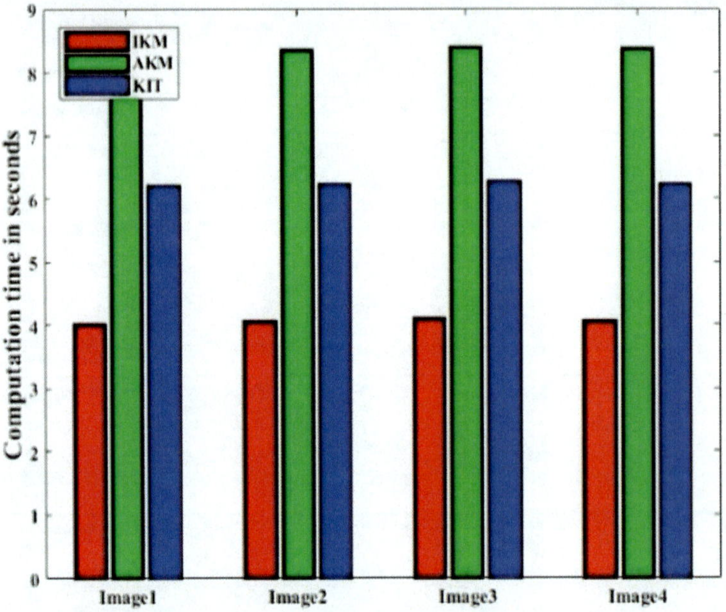

(a) Lv = 7 time

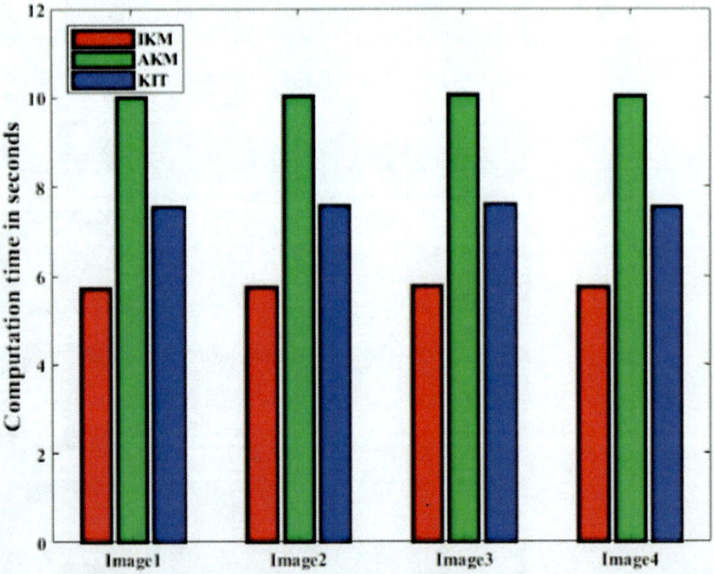

(b) Lv = 11 time

Fig. 3 Box plots and time plots

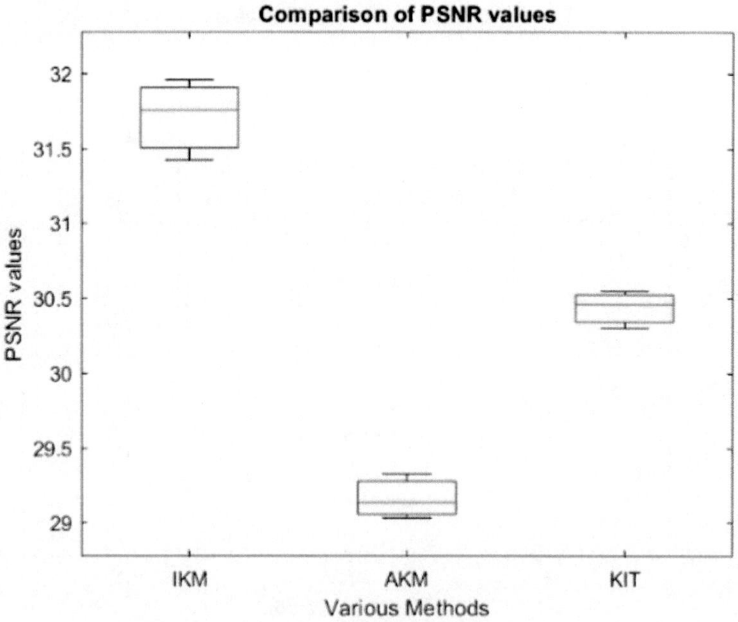

(c) PSNR = 7 box plot

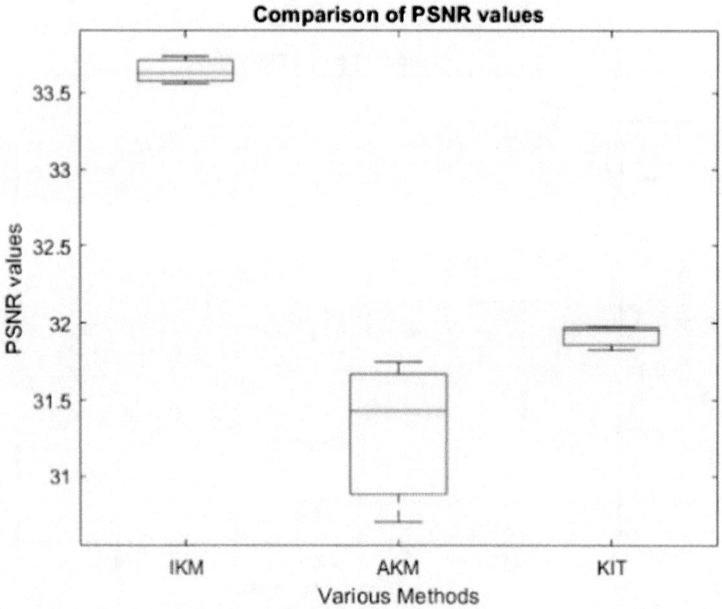

(d) PSNR = 11 box plot

Fig. 3 (continued)

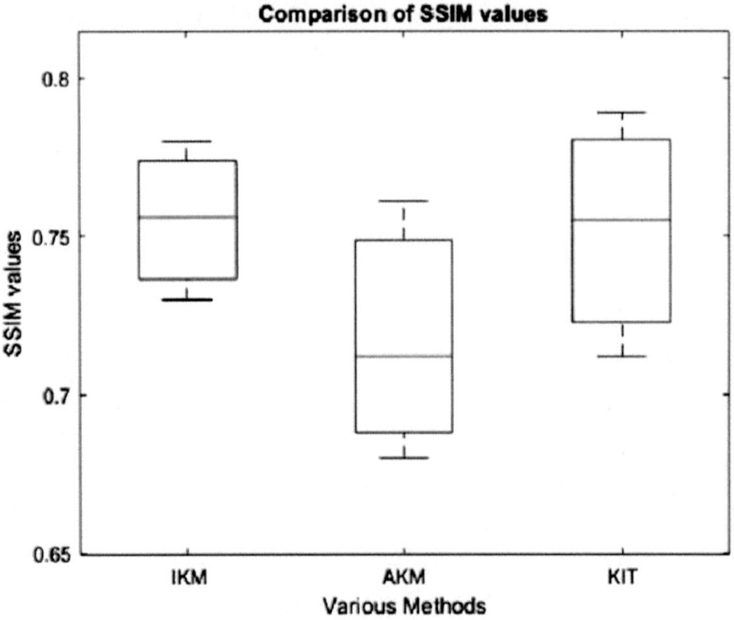

(e) SSIM = 7 box plot

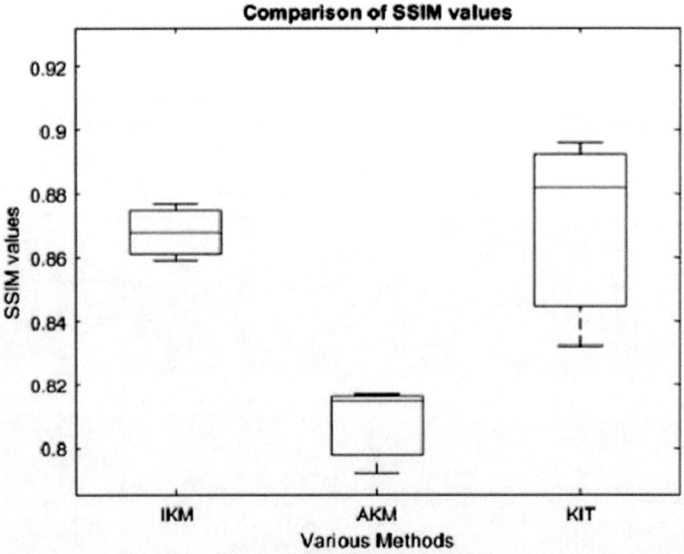

(f) SSIM = 11 box plot

Fig. 3 (continued)

7 Conclusion and Future Work

Image segmentation has been a heavily researched area in recent years. Though many algorithms have been already developed, still it is a huge area of research. This algorithmic approach need not require any user interaction, not even the initial point. Using the K-means algorithm, it identifies the nearby values of the points for clustering the image but cannot always do the process randomly and segments accordingly to distinct images. In this way, machine is set to learn how to segment images and finds out the objects in them. Accuracy is a parameter which is essential to achieve which can be done by more precise and iterative system. And the repetitive initialization can result to more efficient algorithm. It can be improved by incorporating more details about the image. And hence, we have modified the K-means algorithm to have a more precise cluster and segmentation.

References

1. Praveena SM, Vennila I (2010) Optimization fusion approach for image segmentation using K-means algorithm. Int J Comput Appl 2(7):1–9
2. Burney SMA, Tariq H (2014) K-means cluster analysis for image segmentation. Int J Comput Appl 96(4):1–8
3. Milletari F, Navab N, Ahmadi SA (2016) Fully convolutional neural networks for volumetric medical image segmentation. In: International conference on medical image computing and computer assisted interventions, pp 1–11
4. Jumb V, Sohani M, Srivas A (2014) Color image segmentation using K-means clustering and Otsu's adaptive thresholding. Int J Innov Technol Explor Eng 3(9):1–5
5. Dubey SR, Jalal AS (2012) Detection and classification of Apple fruit diseases using complete local binary patterns. Research Gate, pp 2–7
6. Dhanachandra N, Manglem K, Yambem JC Image Segmentation using *K*-means clustering algorithm and subtractive clustering algorithm. Procedia Comput Sci 54
7. Qureshi MN, Ahamad MV (2018) An improved method for image segmentation using K-means clustering with neutrosophic logic. In: Eleventh International multi-conference on information processing-2015 (IMCIP-2015). Procedia Computer Science, pp 534–540
8. Naz S, Majeed H, Irshad H (2010) Image segmentation using fuzzy clustering: a survey. In: 6th International Conference on Emerging Technologies, ICET 2010. IEEE, pp 1–6
9. Dehariya VK, Shrivastava SK, Jain RC (2010) Clustering of image data set using K-means and fuzzy K-means algorithms. In: International conference on computational intelligence and communication networks. IEEE, pp 1–6
10. Hussain HM, Benkrid K, Seker H, Erdogan AT (2011) FPGA implementation of K-means algorithm for bioinformatics application: an accelerated approach to clustering microarray data. In: Conference on adaptive hardware and system AHS 2011. IEEE, pp 1–8
11. Li BN, Chui CK, Chang S, Ong SH (2011) Integrating spatial fuzzy clustering with level set methods for automated medical image segmentation. Comput Biol Med 41(7):1–10
12. Hong Y, Qingling D, Daoliang L, Jianping W (2012) An improved K-means clustering algorithm for fish image segmentation. Math Comput Model 1–9
13. Dubey SR, Dixit P, Singh N, Gupta JP (2013) Infected fruit part detection using K-means clustering segmentation technique. Int J Artif Intell Interact Multimedia 2(2):1–8
14. Chen D, Sain SL, Guo K (2012) Data mining for the online retail industry: a case study of RFM model-based customer segmentation using data mining. Database Market Customer Strategy Manage 19(3):197–208

15. Gong M, Liang Y, Shi J, Ma W, Ma J (2013) Fuzzy C-means clustering with local information and kernel metric for image segmentation. IEEE Trans Image Process 22(2):1–12
16. Vijay J, Subhashini J (2013) An efficient brain tumor detection methodology using K-means clustering algorithm. In: International conference on communication and signal processing on advancing technology for humanity. IEEE, pp 1–5
17. Jose A, Ravi S, Sambath N (2014) Brain tumor segmentation using K-means clustering and Fuzzy C-means algorithms and its area calculation. Int J Innov Res Comput Commun Eng 2(3):1–6
18. IMCIP 2015 (2015) Eleventh International multi-conference on information processing, pp 764–771
19. Abdel-Maksoud E, Elmogy M, Al-Awadi R (2015) Brain tumor segmentation based on a hybrid clustering technique. Egypt Inf J 1–11
20. Singh V, Misra AK (2016) Detection of plant leaf diseases using image segmentation and soft computing techniques. Inf Process Agric 1–26
21. Fränti P, Sieranoja S (2019) How much k-means can be improved by using better initialization and repeats. Pattern Recogn 93(3):95–112
22. Zheng X, Lei Q, Yao R, Gong Y, Yin Q (2018) Image segmentation based on adaptive K-means algorithm. EURASIP J Image Video Process

Chapter 13
Breast Cancer Histopathological Image Classification Using Convolutional Neural Networks

Ankita Adhikari, Ashesh Roy Choudhuri, Debanjana Ghosh, Neela Chattopadhyay, and Rupak Chakraborty

Abstract Nowadays, the classification of medical images has become an essential part of identifying the disease. Among various existing critical diseases, identification of breast cancer has now come up with the topic of investigation. To identify the affected regions of the images, a deep learning-based approach has got wide attention for decades. Convolutional neural networks (CNN) among all deep learning techniques proved their best efficiency in this field. In this paper, one improved CNN-based approach has been proposed to classify the breast cancer images obtainable from the standard **PatchCamelyon (PCam)** benchmark dataset. It is available for free from the website link https://www.kaggle.com/c/histopathologic-cancer-detection/data. In the improved model, various existing layers like convolutional, ReLU, pooling, fully connected have been added as well as modified for better efficiency and efficacy of the algorithm. Further to be added that Adam optimizer has been used here with cross entropy as a loss function. This improved model has been compared with two recent CNN-based approaches applied to medical datasets. The comparative outcomes suggest strong improvements in terms of classification accuracy (probably 2–3%) and computational time of validation loss for both training and testing data over existing models.

Keywords Convolutional neural network · PatchCamelyon data · Adaptive gradient algorithm (AdaGrad) · RMSProp · Validation loss · Validation accuracy

1 Introduction

After non-melanoma skin cancer, breast cancer is the second most common invasive cancer in women, with over 1.5 million women (25% of all women with cancer) survivors of breast cancer worldwide every year [1, 2]. In either the lobules or the breast ducts, cancer develops and may also occur inside one's breast in the fatty tissue

A. Adhikari · A. R. Choudhuri · D. Ghosh · N. Chattopadhyay · R. Chakraborty (✉)
Department of Computer Science and Engineering, Guru Nanak Institute of Technology, Kolkata, India
e-mail: rupak.chakraborty@gnit.ac.in

J. K. Mandal et al. (eds.), *Proceedings of International Conference on Innovations in Software Architecture and Computational Systems*, Studies in Autonomic, Data-driven and Industrial Computing, https://doi.org/10.1007/978-981-16-4301-9_14

or the fibrous connective tissue. It is generally characterized by symptoms ranging from breast pain, red pitted skin over the whole breast, discharge of the bloody nipple, swelling of the breast, erythema of the breast, unexplained changes in the size of the breast, inverted nipple, lump under the arm, and thickening of the skin of the breast. Such signs do not necessarily lead to breast cancer, but they may be a sign of a benign cyst. Breast cancer is divided into two types: invasive, which spreads to other areas of the breast and non-invasive, which stays in the initial tissue. Invasive ductal carcinoma (50–75% of patients), invasive lobular carcinoma (5–15% of patients), and others such as Paget's disease of the nipple, phyllodes tumor, angiosarcoma, make up the remaining patients, are major typical disorders in the two groups listed above [1]. Moreover, in breast cancer pathogenesis, estrogen receptor alpha (ER-alpha) and epidermal growth factor 2 are the two primary molecular targets (ERBB 2 or HER 2). Triple-negative breast cancer arises in a special case in which estrogen receptors are absent from progesterone receptors, and the tumor has no additional HER2 protein on the surface.

Breast cancer diagnosis varies from breast sonograms, magnetic resonance imaging (MRI), biopsy, hormone receptor test, HER2/neu test, and diagnostic mammography. The histopathological diagnosis of breast cancer has reached a larger audience in modern times [3]. Also here, though, the final examination of histopathological breast cancer cells is performed by qualified pathologists physically under the microscope. Computer-Aided Design (CAD) technologies have been developed in recent times to help pathologists identify such cells quicker. But, even the most advanced technologies face complexities in classifying such histopathological pictures, often finding it difficult to decipher the cancerous cells in the tissues more quickly.

There is a regular rise in the contribution of deep learning techniques in medical science [4–6]. A more advanced part of machine learning, consisting of several layers of artificial neural networks, is generally deep learning. Artificial neural networks are a series of linked perceptrons where some function applied to the sum of inputs generates the output. A set in our brains of nodes close to neurons. Moreover, also their working mechanism is similar to that of the nervous system of an ordinary mammal, where the lower-level characteristics (such as lines, curves, edges) learned in the earlier layers converge at the deeper levels to classify the input's higher-level characteristics.

In addition to the many current architectures, convolutional neural networks are an important model that works with visual feedback in this deep learning paradigm. Thanks to its high precision, CNN is used for image detection and identification. This follows a hierarchical model that works to create a network, providing completely interconnected layers where all the neurons are interconnected, and the output is processed. In addition to frequent advances in convolutional neural networks, increasing GPU efficiency, high computational power has made it one of the most advanced deep learning architectures used in medical imaging [7]. It is a specialized network model for deep learning designed to work with two-dimensional image data and can also be applied to one-dimensional and three-dimensional data.

The best thing about CNN is that without any human supervision it extracts and identifies the features from the data automatically. Provided several photographs of cats and dogs, for example, it learns distinctive characteristics on its own for each breed. Another main feature is that deep convolutional networks are versatile and function well on image information so that they are also computationally effective. Special convolution and pooling operations are used, and parameter sharing is carried out. This makes it possible for CNN models to run on any computer, making them universally appealing. This, all in all, sounds like pure sorcery. In order to achieve superhuman precision, we are working with a very strong and efficient model that performs automated feature extraction. CNN models now perform better than humans with image recognition. These days, image classification using pre-trained models is also becoming popular [8, 9], despite not always being a very beneficial medical image classification option [10].

While applying this model, digital photographs of medical ailments are therefore an absolute necessity [4]. Ultrasound (US), magnetic resonance imaging (MRI), positron emission tomography (PET), computed tomography (CT), histology images, X-ray, and many more are the several types of digital imaging used in medical diagnosis [6, 11]. Out of 30.9 million imaging tests, 35% were for advanced diagnostic imaging such as computed tomography (CT), magnetic resonance imaging (MRI), nuclear medicine, and ultrasound during the 15-year study period [12]. In addition, the use of advanced diagnostic imaging increased from 1996 to 2010, where the popularity of CT, MRI, X-ray images has increased by 7.8, 10, and 57%. Another important thing to remember is that compared to an MRI image each image form has varied memory requirements such as a histology image occupies very less memory. The type of image selected as an input to CNN is therefore of critical significance.

This classification was used on the MRI images performed here in [13–15]. Neural networks have been applied to ultrasound images in the following articles [16–18]. This classification was used on the PET images performed here in [19–21]. Histopathological images are used here from the BreakHis dataset [22]. We tried to develop a modified simple convolutional neural network in this paper to solve the problem of classifying and detecting the breast images into normal, benign (non-cancerous abnormality), and malignant (cancer abnormality) categories [23, 24]. Finally, we analyze its effectiveness with the pre-trained models. This is aimed at speeding up the diagnostic process because the conventional method to classify breast cancer takes time, resources, and more manpower by collecting pathological characteristics from qualified experts in this field [3].

This problem is formulated in Sect. 2. Our optimized CNN model is suggested in Sect. 3. The outcomes of our proposed model are discussed in Sect. 4, and a comparison of its success with the other pre-trained models is given. The conclusion of this paper is eventually taken from Sect. 5 and outlines the potential scope.

2 Problem Formulation

Our objective is to identify the presence of lymph node metastases from these 96 × 96 px histopathologic images, and further deeper analysis of these lymph node metastases determines the stage of cancer. The cons of the traditional diagnosis method are that only experts of this domain can be consulted, further increasing their load and demand. Moreover, these metastases being very small in size may get missed easily. Hence, a CNN model handling this work proves effective.

2.1 Convolutional Neural Network

A CNN consists of several layers: convolutional, pooling, fully connected layers, and activation functions like sinh, tanh, ReLU are also used. It takes an image as input then finally classifies it by placing it within a class that has the highest probabilistic score in the last layer [4, 11] (Fig. 1).

Convolution layer. It is the first layer that extracts the low-level features of the input image. It consists of two functions operating on the image: the first one being an array of numeric values of all positions in the input image and the second one being a filter which is also commonly referred to as the kernel which too is a numeric array. The output is the dot product of the above-aforementioned functions. This output is calculated continuously as the filter strides over the input image matrix until it reaches the end. Finally, an activation map is generated, where the filter gets activated, and it identifies lower-level features like an edge, a dot, or a crooked line. If the input image was that of a flower, its simple edges, dots, curved lines would be learnt first by the initial filters. This output would be sent as input in the subsequent layers in order to learn the next more distinct, complicated, higher-level features of the image, finally culminating in learning the entire image. When a filter Kr(a) is slid over an input In(t), the output feature map fm(t) is the following:

$$\mathrm{fm}(t) = (\mathrm{In} * \mathrm{Kr})(t) \tag{1}$$

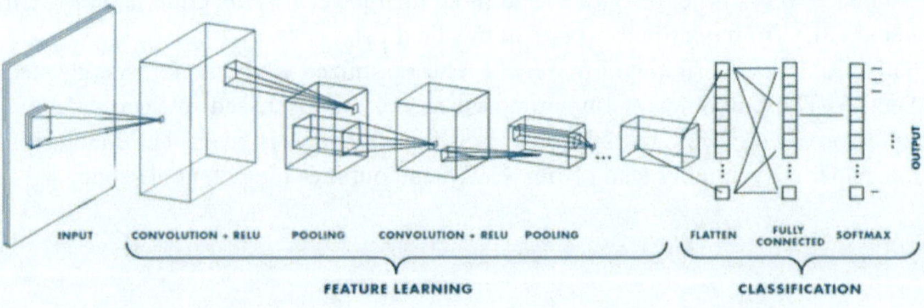

Fig. 1 A simple convolutional neural network

The discrete convolution assuming that t is an integer,
(for 1-D convolution):

$$\mathrm{fm}(t) = \sum_{a} \mathrm{In}(a)\mathrm{Kr}(t - a) \tag{2}$$

(for 2-D convolution):

$$\mathrm{fm}(t) = \sum_{a} \sum_{b} \mathrm{In}(a, b)\mathrm{Kr}(m - a, n - b) \tag{3}$$

Rectified linear unit layer (ReLU). It is a piecewise linear function that gives zero as the output whenever the input is negative and gives the input as the output whenever the input is positive. ReLU fastens the entire training and avoids the vanishing gradient problem. It can be represented by:

$$\mathrm{fn}(x) = \max(0, x) \tag{4}$$

x is the input to the neuron. Compared to other activation functions, ReLU gives better performance results and is used more in most neural networks.

Pooling layer. It is usually present between the convolution and rectified linear unit layers in order to decrease the number of calculated parameters. Max-pooling returns the highest value among all the input values within a filter over the image discarding the remaining values. Similarly, average pooling returns the average of all input values. Acting as a dimensionality reducer and noise suppressant, max-pooling works more effective than average pooling.

Fully connected layer. The output of the layers before becomes the input for the fully connected layer (FC), that is our feature map matrix; then, it is flattened to finally convert them into a single vector which in turn acts as the next stage input. The objective of an FC layer is to take the results of the convolution/pooling process and uses them to classify the image into a class label. This layer undergoes backpropagation to estimate the most accurate weights. Each neuron prioritizes the most appropriate label using the weights they receive. Finally, the neurons conduct "voting" for each label to judge the winner and come to a classification decision [25]. The activation functions help in the classification process.

3 Proposed Model

3.1 Our CNN Architecture

See Table 1.

Table 1 Our CNN model architecture

Layer(type)	Output shape	Param
conv2d_9 (Conv2D)	(None, 94, 94, 32)	896
conv2d_10 (Conv2D)	(None, 92, 92, 32)	9248
conv2d_11 (Conv2D)	(None, 90, 90, 32)	9248
max_pooling2d_3	(MaxPooling2 (None, 45, 45, 32)	0
dropout_4 (Dropout)	(None, 45, 45, 32)	0
conv2d_12 (Conv2D)	(None, 43, 43, 64)	18,496
conv2d_13 (Conv2D)	(None, 41, 41, 64)	36,928
conv2d_14 (Conv2D)	(None, 39, 39, 64)	36,928
max_pooling2d_4	(MaxPooling2 (None, 19, 19, 64))	0
dropout_5 (Dropout)	(None, 19, 19, 64)	0
conv2d_15 (Conv2D)	(None, 17, 17, 128)	73,856
conv2d_16 (Conv2D)	(None, 15, 15, 128)	147,584
conv2d_17 (Conv2D)	(None, 13, 13, 128)	147,584
max_pooling2d_5	(MaxPooling2 (None, 6, 6, 128))	0
dropout_6 (Dropout)	(None, 6, 6, 128)	0
flatten_1 (Flatten)	(None, 4608)	0
dense_2 (Dense)	(None, 256)	1,179,904
dropout_7 (Dropout)	(None, 256)	0
dense_3 (Dense)	(None, 2)	514

Total params: 1,661,186; Trainable params: 1,661,186; Non-trainable params: 0

3.2 Discussion

3.2.1 Our CNN Model

Input layer. The work done by input layer involves loading of input prior to producing output used to feed convolutional layers, which can involve future scaling application. In the present case, images and the parameters are inputs and image dimension (94 $\times$ 94 pixels), respectively, the number of channels being 3 for RGB.

Convolutional layers. There are 9 convolutional layers in this model in sets of 3 each set followed by a pooling layer. The size of receptive kernels is 3 $\times$ 3. The first three convolutional layers learn 32 filters each. The middle three layers learn 64 filters and the final 3 filters 128.

Pooling layers. To facilitate downsampling of the input's spatial dimension, pooling layers are used. Each convolutional layer is preceded by one convolutional layer.

They are used to design max-pooling filter of 2 rows × 2 columns and a stride of 2 rows × 2 columns. The most common max operations are used by the pooling layers over the receptive field.

ReLU layers. ReLU activation function is used in these layers, hence, if the input value is greater than 0, the value is the direct output or if the input is less than 0, then the output is 0.

Dropout. During the time of training with a frequency of rate at each step, the dropout layer sets input units to 0, thereby preventing overfitting. After every pooling layer, dropout is being used.

Flatten Class. The function layer serves as a utility layer that flattens an input having a shape of n * c * h * w, resulting in a simple vector output having a shape of n * (c * h * w).

Fully connected layers. Here, we have 2 classes, so the final FC layer has 2 nodes. We implement fully connected layers using the dense function.

Softmax. The output classification layer is preceded by a softmax layer. Output layer which is fully connected having a softmax activation is dependent on the number of classes in the classification problem, i.e., for binary classification problems, and there are 2 output filters.

Optimizer. We have trained our model over 20 epochs using Adam optimizer with a cross-entropy loss function. It has an initial learning rate of 0.0001 and validation patience 2. An epoch describes the number of times the CNN algorithm is applied over the entire training dataset. Adam is an adaptive learning rate optimization algorithm which means it calculates individual learning rates for different parameters, and it has been designed specifically for training deep neural networks. Here, the estimated calculations made during the first and second moments of the gradient are utilized for adapting the learning rate for each weight of the neural network. Adam can be considered as the child of RMSProp and AdaGrad (Fig. 2).

4 Experimental Results

4.1 Dataset

In this paper, the PatchCamelyon (PCam) benchmark dataset is used. It can be found free of charge under the website link https://www.kaggle.com/c/histopathologic-cancer-detection/data. The original PCam dataset is available for free at https://github.com/basveeling/pcam link on the website. Due to its probabilistic sampling, it includes duplicate images, but the version shown on Kaggle does not contain duplicates. It is a new challenging dataset of image classification with 96 × 96 px color images extracted from histopathologic scans of sections of lymph nodes. With each

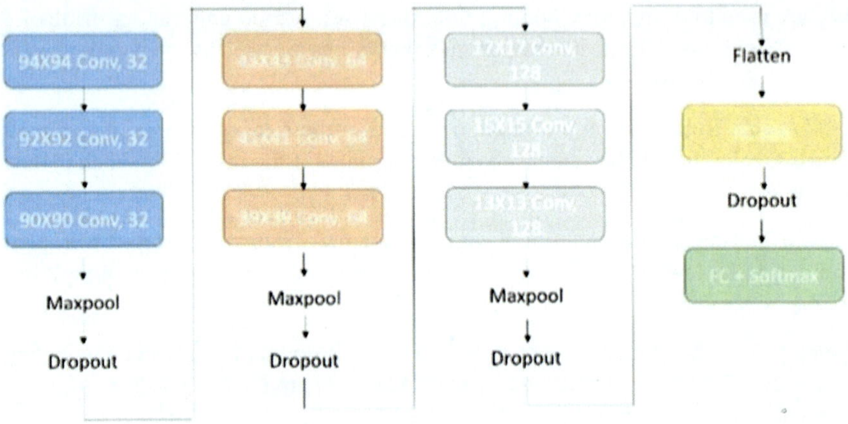

Fig. 2 The proposed convolutional neural network of our model with the specified layers

image demonstrating the presence of metastatic tissue, a binary label is provided. The training, testing, validation split in the dataset has been described in the Table 2.

This dataset for PCam was taken from the Camelyon16 challenge [22] (Fig. 3). Four hundred whole-slide or histopathological images of sentinel lymph nodes-stained hematoxylin and eosin (H&E) are included in this dataset. All data in this challenge has been collected from two separate datasets obtained from the Utrecht University Medical Center (Utrecht, the Netherlands) and the Radboud University Medical Center (Nijmegen, The Netherlands). Using a 40 × goal, they were loaded

Table 2 Split in the dataset made in the following way

Split type	Images
Train	262,144
Test	32,768
Validation	32,768

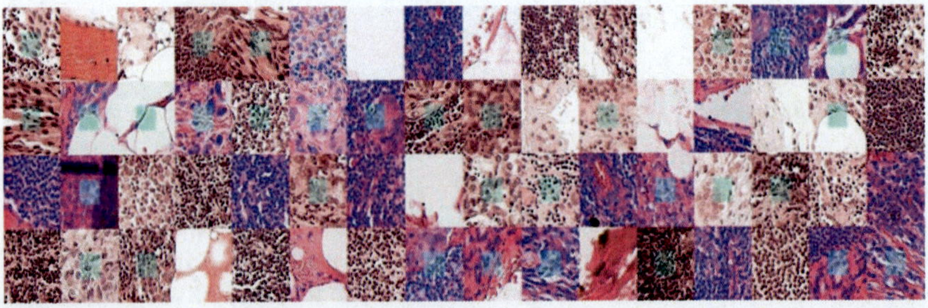

Fig. 3 Stained PCam histopathological images where malignant tissue of breast cancer is denoted by the stained green boxes. Image used from the following link: https://github.com/basveeling/pcam/blob/master/pcam.jpg

Table 3 PCam benchmark dataset image distribution

Type	Number
Benign	89,117
Malignant	130,908

at 2 separate centers. It was further under-sampled to 10 × to expand the field of view. Table 3 displays the number of benign or positive images and the number of negative or metastasized or malignant images in the dataset.

4.2 Simulation Results Analysis

Table 2 describes the way the data is split. Like any other classification model, we first train our entire dataset over the model, then test it, then use the validation set to perform fine-tuning, our final goal being to classify the malignant images and the benign ones. The proposed model was implemented using the TensorFlow framework in Jupyter Notebook on a Windows OS machine with AMD Ryzen 5 2400 processor, 12 Gb Ram, and NVIDIA 1050 ti GPU. Figure 4 is the confusion matrix of our proposed model that further describes the performance of the model. The (0,0) and (1,1) quadrants show the true negative and true positives, respectively, which is much

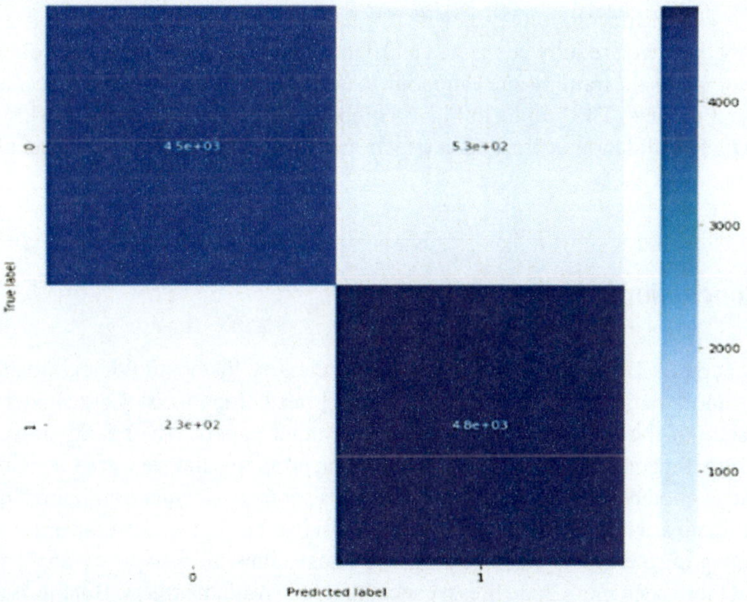

Fig. 4 Confusion matrix of the proposed model

Table 4 Image classification analysis between the different models

CNN model	Training accuracy (%)	Testing accuracy (%)	Average accuracy (%)	Training time (s)	Test time (s)
Proposed CNN	93.88	92.40	93.14	55,390	858
CNN [19]	92.18	91.40	91.79	55,401	860
CNN [26]	92.11	90.73	91.42	55,394	859

higher than (represented by darker shades) the number of false negatives (0,1) and false positives (1,0).

Table 4 provides the average classification accuracy of our proposed CNN model, and a comparison is made with two other models [19] and [26] that have been implemented on our dataset. All these models described by the authors in their corresponding papers also followed the train-test-validate mechanism. As described in the first entry of Table 4, the training and testing accuracies of our model are 93.88% and 92.14%, the average classification accuracy is 93.14%, better than the other two models CNN [19] and CNN [26] with an average classification accuracy of 91.79% and 91.42%, respectively.

CNN [19] has 5 convolutional layers, and the size of the kernel in each layer is 2 × 2, followed by a 27 × 27 pooling layer. The leaky ReLu activation function has been used for activating every convolutional layer while standard ReLu has been used in the first two layers and softmax on the last layer of the dense part of the network. CNN [26] has 3 convolution layers with 5 × 5 filter, 3 pooling layers with 2 × 2 filter where the last two are fully connected and 3 tanh activation functions. The following Fig. 5 compares the training and validation loss and training and validation accuracies of these 3 models. The graphs in Fig. 5 compare the loss and accuracy during both training and validation against each epoch, for all the three models represented by different colors.

5 Conclusion and Future Scope

In this paper, a CNN model was proposed to classify histopathological images of breast cancer patients. Finally, we provided a brief comparison of our model with some state-of-the-art CNN architectures from other papers. The results prove that our design is a good solution to this classification problem, and we were successful in attaining a good benchmark performance. In this paper, an adequate amount of breast cancer images were processed. Yet we also discovered that we had our limitations while adapting the CNN models to the given dataset. It would also prove more fruitful and had there been more good breast cancer datasets available freely. But this is a feat that might require some time to achieve, since collecting private images of patients' ailments come under personal health security norms and the cost of collection of

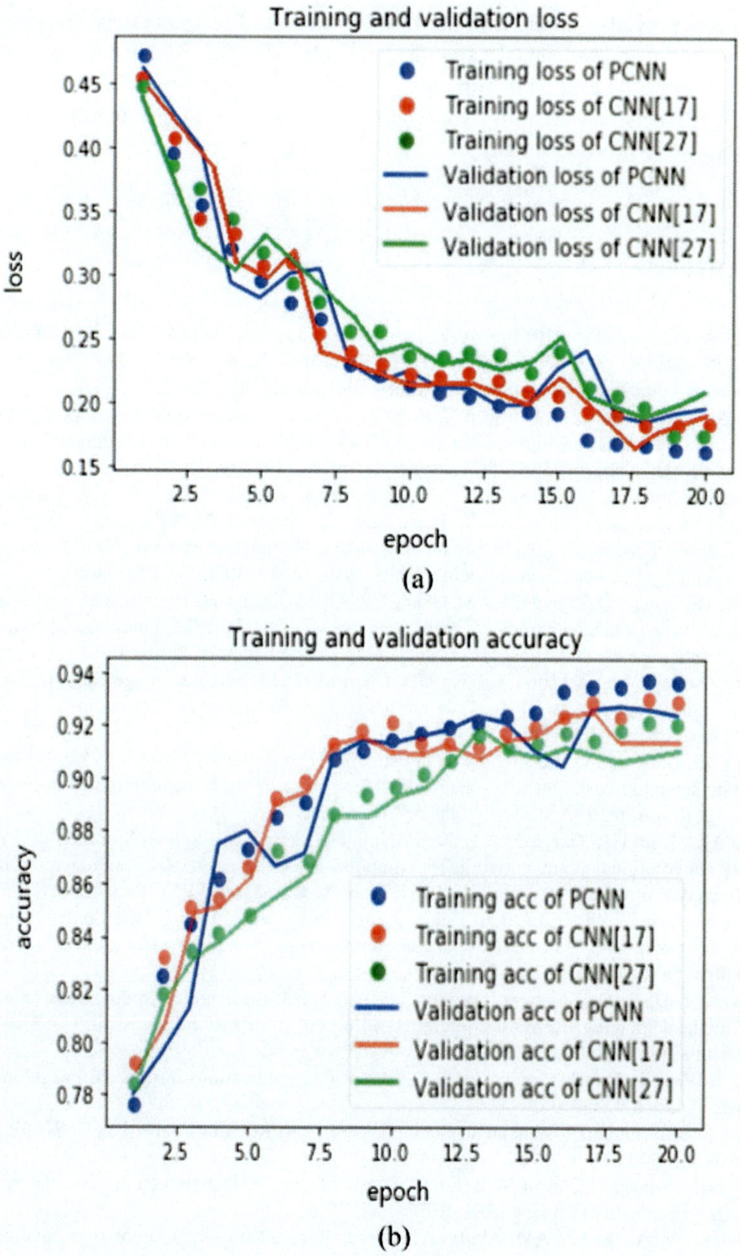

Fig. 5 **a** Training and validation loss comparison chart and **b** Training and validation accuracy comparison chart of the three models where blue curves denote our proposed CNN model, red curves denote CNN [19], and green curves denote CNN [26] of Table 4

such images may also be high. In the future, more modifications may be applied to this CNN model and we can try expanding it to classify other image datasets like that of the lung, heart, kidney, skin ailments. We may also use different medical imagery like PET, CT scans, MRI to do so, keeping in mind the unique features of every image type.

References

1. Waks AG, Winer EP (2019) Breast cancer treatment a review. JAMA 321(3):288–300
2. Sun YS, Zhao Z, Yang ZN, Xu F, Lu HJ, Zhu ZY, Shi W, Jiang J, Yao PP, Zhu HP (2017) Risk factors and preventions of breast cancer. Int J Biol Sci 13(11):1387–1397
3. Spanhol FA, Oliveira LS, Petitjean C, Heutte L (2016) Breast cancer histopathological image classification using convolutional neural networks. International Joint Conference on Neural Networks (IJCNN), pp 2560–2567. https://doi.org/10.1109/IJCNN.2016.7727519
4. Smith-Bindman R, Miglioretti DL, Johnson E, Lee C, Feigelson HS, Flynn M, Greenlee RT, Kruger RL, Hornbrook MC, Roblin D, Solberg LI, Vanneman N, Weinmann S, Williams AE (2012) Use of diagnostic imaging studies and associated radiation exposure for patients enrolled in large integrated health care systems, 1996–2010. JAMA 307(22):2400–2409
5. Xu Y, Mo T, Feng Q, Zhong P, Lai M, Chang EIC (2014) Deep learning of feature representation with multiple instance learning for medical image analysis. In: IEEE International conference on acoustic, speech and signal processing (ICASSP). IEEE, pp 1626–1630
6. Li Q, Cai W, Wang X, Zhou Y, Feng DD, Chen M (2014) Medical image classification with convolutional neural network. In: 13th International conference on control, automation, robotics & vision. IEEE, pp 844–848
7. Raj RJS, Shobana SJ, Pustokhina IV, Pustokhin DA, Gupta D, Shankar K (2020) Optimal feature selection-based medical image classification using deep learning model in internet of medical things. IEEE Access 8:58006–58017
8. Shin HC, Roth HR, Gao M, Lu L, Xu Z, Nogues I, Yao J, Mollura D, Summers RM (2016) Deep convolutional neural networks for computer-aided detection: CNN architectures, dataset characteristics and transfer learning. IEEE Trans Med Imaging 35(5):1285–1298
9. Saito K, Zhao Y, Zhong J (2019) Heart diseases image classification based on convolutional neural network. In: 2019 International conference on computational science and computational intelligence (CSCI). IEEE, pp 930–935
10. Raghu M, Zhang C, Kleinberg J, Bengio S (2019) Transfusion: understanding transfer learning for medical imaging. In: 33rd conference on neural information processing systems. arXiv preprint arXiv:1902.07208
11. Ker J, Wang L, Rao J, Lim T (2017) Deep learning applications in medical image analysis. IEEE Access 6:9375–9389
12. Veeling BS, Linmans J, Winkens J, Cohen T, Welling M Rotation equivariant CNNs for digital pathology. arXiv:1806.03962
13. Farooq A, Anwar SE, Awais M, Rehman S (2017) A deep CNN based multi-class classification of Alzheimer's disease using MRI. IEEE
14. Chen L, Wu Y, Dsouza AM, Abidin AZ, Wismuller A, Xu C MRI tumor segmentation with densely connected 3D CNN. arXiv:1802.02427v2. Accessed 9 Feb 2018
15. Zou L, Zheng J, Miao C, Mckeown MJ, Wang ZJ (2017) 3D CNN based automatic diagnosis of attention deficit hyperactivity disorder using functional and structural MRI. IEEE Access 5:23626–23636
16. Andersen JKH, Pedersen JS, Laursen MS, Holtz K, Grauslund J, Savarimuthu TR, Just SA (2019) Neural networks for automatic scoring of arthritis disease activity on ultrasound images. RMD Open 5

17. Biswas M, Kuppili V, Edla DR, Suri HS, Saba L, Marinhoe RT, Sanches JG, Suri JS (2018) Symptosis: a liver ultrasound tissue characterization and risk stratification in optimized deep learning paradigm. Comput Methods Programs Biomed 155:165–177
18. Hao PY, Xu ZY, Tian SY, Wu FL, Chen W, Wu J, Lu XN (2019) Texture branch network for chronic kidney disease screening based on ultrasound images. FITEE 1–10
19. Yonekura A, Kawanaka H, Prasath VB, Aronow BJ, Takase H (2017) Improving the generalization of disease stage classification with Deep CNN for Glioma histopathological images. IEEE Int Conf Bioinf Biomed (BIBM) 17:1222–1226
20. Wang X, Teng P, Lo P, Banola A, Kim G, Abtin F, Goldin J, Brown M (2018) High throughput lung and lobar segmentation by 2D and 3D CNN on chest CT with diffuse lung disease. Springer Nature Switzerland 11040:202–214
21. Xu X, Jiang X, Ma C, Du P, Li X, Lv S, Yu S, Yu L, Ni L, Ni Q, Chen Y, Su J, Lang G, Li Y, Zhao H, Liu J, Xu K, Ruan L, Sheng J, Qiu Y, Wu W, Liang T, Li L (2020) A deep learning system to screen novel coronavirus disease 2019 Pneumonia. Journal Pre-proofs 1–12
22. Bejnordi E et al (2017) Diagnostic assessment of deep learning algorithms for detection of lymph node metastases in women with breast cancer. JAMA 318(22):2199–2210
23. Bardou D, Zhang K, Ahmad SM (2018) Classification of breast cancer based on histology images using convolutional neural networks. IEEE Access 20:24680–24693
24. Bejnordi BE, Veta M, van Diest PJ, van Ginneken B, Karssemeijer N, Litjens G, van der Laak JA, Hermsen M, Manson QF, Balkenhol M et al (2017) Diagnostic assessment of deep learning algorithms for detection of lymph node metastases in women with breast cancer. JAMA 318(22):2199–2210
25. Yamashita R, Nishio M, Do RKG, Togashi K (2018) Convolutional neural networks: an overview and application in radiology. Insights Imaging 9:611–629
26. Anthimopoulos M, Christodoulidis S, Ebner L, Christe A, Mougiakakou S (2016) Lung pattern classification for interstitial lung diseases using deep convolutional neural network. IEEE Trans Med Imaging 35(5):1207–1216

Chapter 14
A Framework for Predicting Placement of a Graduate Using Machine Learning Techniques

Amrut Ranjan Jena, Subhajit Pati, Snehashis Chakraborty, Soumik Sarkar, Subarna Guin, Sourav Mallick, and Santanu Kumar Sen

Abstract Campus placement carries a great significance for all the students and educational institutes. Nowadays, students give special attention to past placement records while selecting an institution for their admission. Hence, the institutions attempt to improve their graduate job appointment activities. The aim of this work is to evaluate past students' academic records, and forecast placement probability of existing students. This placement predictor model takes different parameters those can be used to analyze the skill level of the student. While some parameters are taken from the institute level data, others are obtained from tests records conducted by the placement itself. Combining these data points, the model predicts, whether a student will be placed or not. Therefore, a model has been proposed to predict the placement possibilities with the help of machine learning algorithms. A framework is designed with the help of eight machine learning algorithms over the collected dataset, and the accuracy of the model is checked through these algorithms. Besides, random forest algorithm gives better performance among all the eight machine learning techniques.

Keywords Machine learning techniques · 5-folds cross-validations · Accuracy · Campus placement · Attribute selection

1 Introduction

In present days the educational establishments grow rapidly. The prime goal of all higher educational institutes focuses on student's placement activities. At present, the major challenge of the institutions faces, how to uplift the student job appointments. The effective ways to improve the placement is through quality teaching learning methods and advanced trainings as per industry requirements [1]. Machine learning helps to extract information from a database that resides at intervals of the institutions records exploitation. The data needed for developing the system consists passed students data. To train the model for rule identification and to test the model for

A. R. Jena (✉) · S. Pati · S. Chakraborty · S. Sarkar · S. Guin · S. Mallick · S. K. Sen
Guru Nanak Institute of Technology, Kolkata, India
e-mail: amrutranjan.jena@gnit.ac.in

J. K. Mandal et al. (eds.), *Proceedings of International Conference on Innovations in Software Architecture and Computational Systems*, Studies in Autonomic, Data-driven and Industrial Computing, https://doi.org/10.1007/978-981-16-4301-9_15

prediction, this data is analyzed. The prediction of job appointments that students are supposed to get helps them to put extra effort to get a job through campus drive. This model helps all the stakeholders of an institution to provide better care to increase the number of job appointments. Often the placement scenario is a major attribute for projecting the status of an institution. Therefore, the placement prediction model plays major role in educational industry. In this work, eight machine learning (ML) techniques are applied over the collected dataset to predict the placement chances of a graduate. This placement prediction model will be productive for both the students as well as the institution for maintain a good placement record. The ML techniques include random forest, decision tree, k-nearest neighbor (KNN), adaptive gradient boosting (ADA), logistic regression, gradient boost, extreme gradient boosting (XG), and support vector machine (SVM) learning technique, and described below.

Logistic Regression: It projects the probabilistic worth exploitation provision that operates in a given applier, which can make up one in all two classes like accept admission or rejects admission. The provision regression facilitates an easy model fitting to explain the linking among variables, and freelance variables. The attributes of the provision operated to optimize exploitation probability [2].

Random Forest: It is a well-known classifier used to support the mix of tree predictors to pretend that a tree is individually formed within the group. Once some ranges of trees are produced, every tree cast a vote to find a brand-new information so that the maximum prediction wins. A number of attributes which will be adjusted within this technique are: the tree counts within the group, range of options for separating every node resulting minimum samples requirement to form the split [3].

Decision trees: It employs a tree-based construction for representing different attainable calls associated with each path outcomes. A call tree begins from the root node and the branches come out to multiple attainable call nodes based on ranking options for decreasing the randomness of the sub dataset. The call rules are represented by the edge for culminating the leaves to denote the end result of the learning. The attributes measured in call trees embody, however, aren't restricted to the tree depth [4].

K-Nearest Neighbors: This formula applies the concept of distance among the elements and predicts on the idea that elements having similarities are close to each other. For a specified associate, the unseen observation is computed based on the neighbor properties. The hidden information employs the category according to the majority prediction supported by its K-nearest neighbors [5].

Gradient Boosting: The models do not show better prediction result can be boosted using this method. Besides, the models are boosted by ensemble methodology so that a model having less prediction accuracy approaches toward powerful prediction model [6].

Support Vector Machines: In this method, it associates formula for separating binary categories with associate best hyperplane to increase the boundary of the information. It explores for a call point where it acts like a remote attainable of any information divided by the two categories. The classifier boundary is the distance from the choice point to the nearest information, and these positions are referred as support vectors [7].

AdaBoost formula: This could be wont to boost the performance of any ML algorithm. This has become the strong methods which may build predictions supported an oversized quantity of knowledge. It has become thus standard in recent times that the appliance of machine learning will be found in our daily activities. A typical example of its obtaining suggestions for merchandise whereas searching online supported the past things bought by the client. Machine learning, typically mentioned as prophetical analysis or prophetical modeling, will be outlined because the ability of computers to be told while not being programmed expressly. It uses programmed algorithms to research input file to predict output inside a suitable vary [8].

XG Boost: conjointly called the intense Gradient boost may be a machine learning formula that's used for the implementation of gradient boosting call trees. Why calls trees once we quote unstructured information just like the pictures, unstructured text information, etc., the artificial neural networks model appears to reside at the highest once we attempt to predict. Whereas once we quote structured/semi-structured information, call trees are presently the simplest. XG Boost is primarily designed for up the speed and performance of machine learning models greatly and it served the aim okay [9].

2 Proposed Method

To design the proposed model, the data are collected from Kaggle web resource. For better understanding, a sample dataset is projected in Table 1.

Further, the collected data is analyzed by exploratory data analysis technique. A statistical summary of the dataset is prepared with the help of Pandas [10, 11]. Besides, the separation of numerical features and categorical features are identified from the dataset [12]. It includes 5 numerical features and 8 categorical features. All the features are converted into numerical data before training. The aim of this work refers to forecast placement chances of a graduate by using ML techniques. Eight ML techniques are applied to train the proposed model. These techniques are such as random forest, decision tree, k-nearest neighbor (KNN), adaptive gradient boosting (ADA), logistic regression, gradient boost, extreme gradient boosting (XG), and support vector machine (SVM) learning technique. However, the random forest method predicts better result as compared to other seven machine learning techniques as mentioned above. Thus, the final model is designed with the help of random forest technique as its training algorithm. The proposed model is depicted in Fig. 1. The dataset is dived into training set and testing set. In the training part, it contains 70% of the data to train the model. The rest 30% data is passed through the testing phase to check accuracy of the model. The model accuracy is checked by using 5-folds cross-validation method by computing the mean squared error over all folds [13].

Table 1 Sample student dataset

Sl. No.	Gender	Ssc_p	Ssc_b	Hsc_p	Hsc_b	Hsc_s	Degree_p	Degree_t	workex	Etest_p	Specialization	Mba_p	Status
1	M	67.00	Other	91.00	Other	Commerce	55.00	Sci & Tech	No	55.00	Mkt & HR	58.80	Placed
2	M	79.33	Central	78.33	Other	Science	77.48	Sci & Tech	Yes	86.50	Mkt & Fin	66.28	Placed
3	M	65.00	Central	68.00	Other	Arts	64.00	Comm & Mgmt	No	75.00	Mkt & Fin	57.80	Placed
4	M	56.00	Central	52.00	Central	Science	52.00	Sci & Tech	No	66.00	Mkt & HR	59.43	Not placed
5	M	85.80	Central	73.60	Central	Commerce	73.00	Comm & Mgmt	No	96.80	Mkt & Fin	55.50	Placed
6	M	55.00	Other	49.80	Other	Science	67.25	Sci & Tech	Yes	55.00	Mkt & Fin	51.85	Not placed
7	F	46.00	Other	49.20	Other	Commerce	79.00	Comm & Mgmt	No	74.28	Mkt & Fin	53.29	Not placed
8	M	82.00	Central	64.00	Central	Science	66.00	Sci & Tech	Yes	67.00	Mkt & Fin	62.14	Placed
9	M	73.00	Central	79.00	Central	Commerce	72.00	Comm & Mgmt	No	91.34	Mkt & Fin	61.29	Placed
10	M	83.00	Central	70.00	Central	Commerce	61.00	Comm & Mgmt	No	54.00	Mkt & Fin	52.21	Not placed

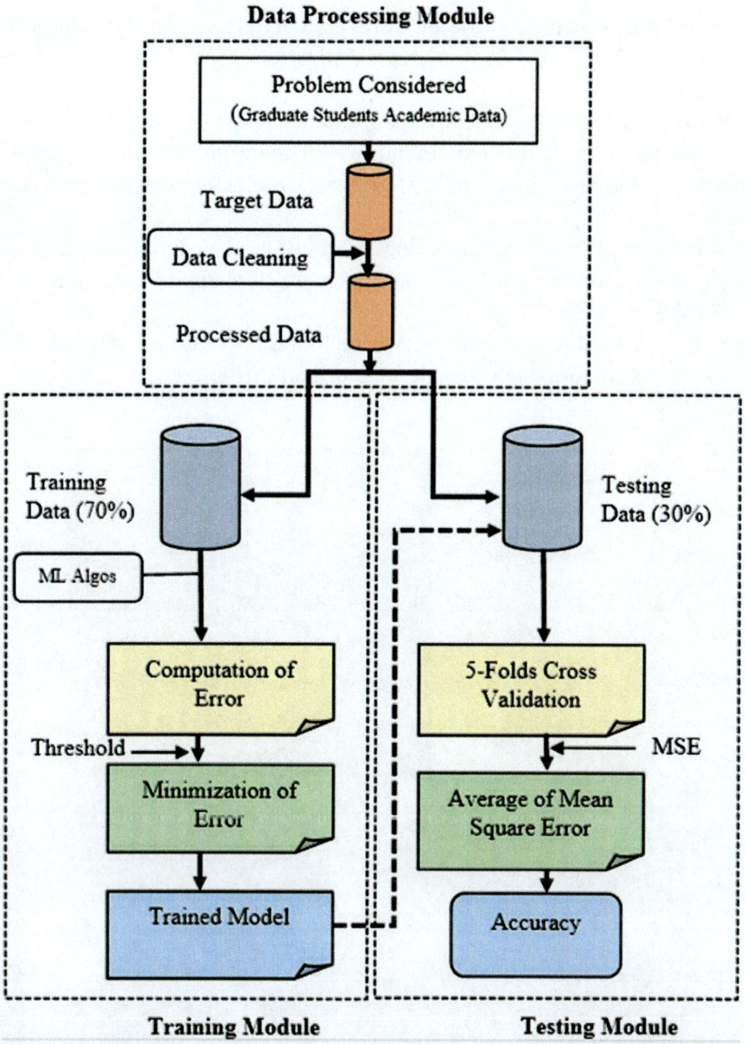

Fig. 1 Proposed model

3 Result Analysis

The model performance is evaluated using random forest, decision tree, ADA, gradient boost, XG, KNN, logistic regression, and SVM. 5-folds cross-validation technique is used to measure the accuracy of the model over above-mentioned algorithms.

For feature selection, three different tests are conducted over the dataset, which includes heat map test, ANOVA test, and chi-square test. All the tests are described in this section.

A heat map is plotted to project the correlation among the attributes. The heat map describes the relevant relation among two attributes to explain the feature. The multipolarity is removed with the help of heat map. For better visualization, it shown in Fig. 2.

In ANOVA test, the probability value (*p*-value) is considered to determine the importance of the attribute toward prediction result. It is applied over the collected dataset and shown in Fig. 3.

Further chi-squared test is applied over the dataset to find the attributes having high impact for predicting the result. It is described in Table 2.

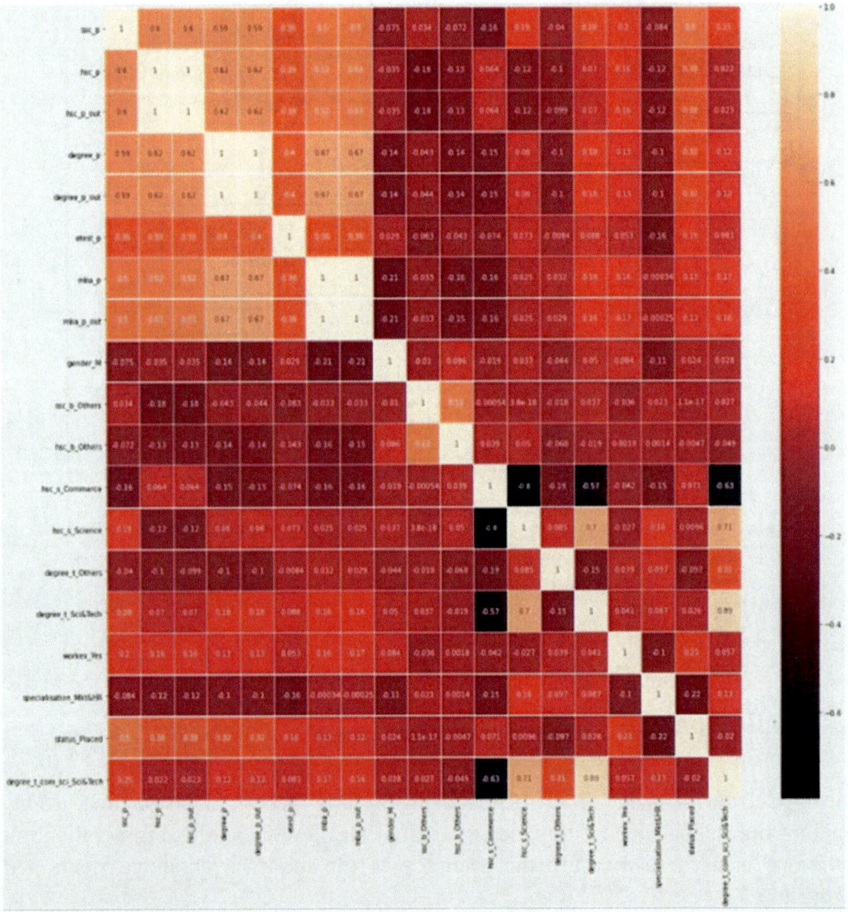

Fig. 2 Heat map test result

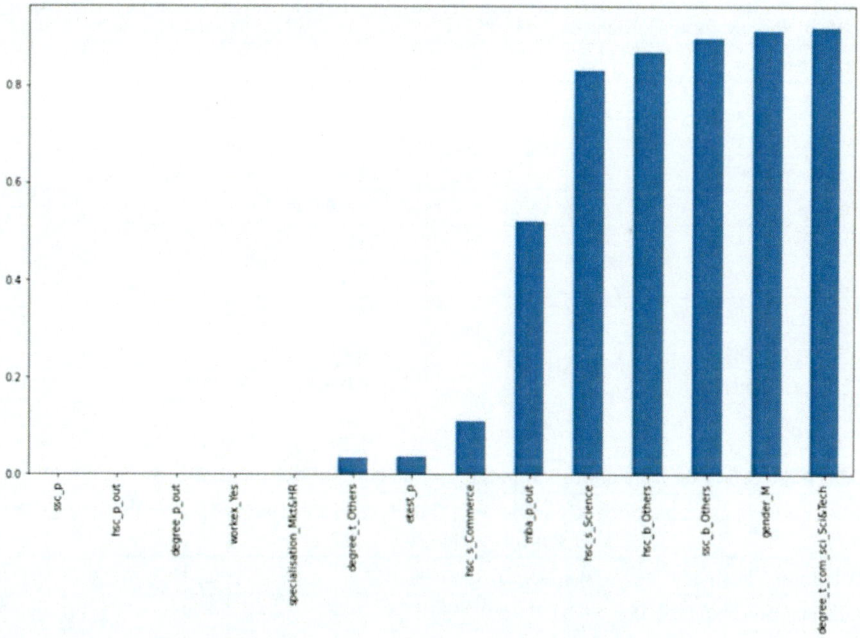

Fig. 3 ANOVA test result

Table 2 Chi-squared test result

Feature name	Corresponding *P*-value
Ssc_p	5.88522955e−20
Hsc_p	4.82587423e−03
Degree_p	1.14928428e−11
Etest_p	1.06254810e−08
Mba_p	3.19031057e−02
Hsc_p_out	6.74255027e−01
Degree_p_out	1.00000000e+00
Mba_p_out	9.34912531e−01
Gender_M	2.21189731e−01
Ssc_b_other	8.69000565e−01
Hsc_b_other	9.24110346e−02
Hsc_s_commerce	2.77393071e−04
Hsc_s_science	1.44701390e−04
Degree_t_other	7.34221002e−01

Table 3 *P* score for the attributes

Feature name	Corresponding *P*-value
Ssc_p	149.020128
Hsc_p_out	88.642854
Degree_p_out	42.296819
Etest_p	17.432197
Workex_Yes	8.039130
Specialization_Mkt & HR	7.722628
Mba_p_out	6.856404
Degree_t_Others	2.700000
Hsc_s_Commerce	0.797872

Table 4 Accuracy measure through 5-folds cross-validation of student dataset

Machine learning techniques	Accuracies over 5-folds cross-validation of student dataset	Mean of the accuracies
AdaBoost Classifier	0.8, 0.83, 0.75, 0.78, 0.77	0.79
Gradient Boosting Classifier	0.83, 0.82, 0.75, 0.85, 0.73	0.79
Random Forest Classifier	0.87, 0.83, 0.78, 0.85, 0.72	0.81
Multinominal INB	0.77, 0.8, 0.77, 0.77, 0.75	0.77
Decision Tree Classifier	0.72, 0.8, 0.68, 0.77, 0.65	0.72
XG Boost	0.85, 0.82, 0.73, 0.83, 0.7	0.79
SVC	0.67, 0.75, 0.77, 0.88, 0.75	0.76
Logistic Regression	0.72, 0.73, 0.77, 0.87, 0.8	0.78
KNN Classifier	0.65, 0.68, 0.75, 0.85, 0.75	0.74

From Table 2, it is observed that few attributes have very less impact on predicting the result. Therefore Table 3 shows the high impact attributes for computation.

Finally, the model accuracy is calculated through 5-folds cross-validation method and shown in Table 4. From Table 4, it is clear that the model accuracy of random forest classifier is 0.81 (81%), and highest compared to other seven classifiers.

4 Comparative Analysis

This section narrates the comparative analysis of this model. The eight algorithms of this framework are applied over the dataset of early stage diabetes risk prediction [14]. The results computed by the model is shown in Table 5.

From Table 5, it is noticed that the random forest technique gives better results as compared to other seven machine learning techniques. Thus, it is proved from the

Table 5 Accuracy measure through 5-folds cross-validation for diabetes dataset

Machine learning techniques	Accuracies over 5-folds cross-validation of diabetes dataset	Mean of the accuracies
AdaBoost Classifier	0.82, 0.76, 0.75, 0.73, 0.75	0.76
Gradient Boosting Classifier	0.82, 0.78, 0.81, 0.77, 0.78	0.79
Random Forest Classifier	0.85, 0.82, 0.81, 0.80, 0.81	0.81
Multinominal INB	0.84, 0.81, 0.81, 0.76, 0.77	0.79
Decision Tree Classifier	0.85, 0.72, 0.78, 0.74, 0.79	0.78
XG Boost	0.81, 0.76, 0.78, 0.76, 0.78	0.78
SVC	0.79, 0.72, 0.72, 0.75, 0.76	0.75
Logistic Regression	0.82, 0.79, 0.81, 0.79, 0.76	0.79
KNN Classifier	0.82, 0.76, 0.75, 0.73, 0.75	0.76

comparative study that, the random forest technique gives better results as compared to other seven machine learning techniques in two different datasets. In both the cases, it gives similar accuracy of 81%.

5 Conclusion

In present days getting a job is very important for a graduate. To get a job, a graduate need to prepare itself as per market scenario. Therefore, institutions are preparing students as per their capabilities. In this paper, the placement prediction model is proposed by applying machine learning techniques over student's academic performance records. From the result analysis and comparative study, it is found that random forest technique gives maximum accuracy of 81% as compared to other seven machine learning techniques. Therefore, random forest technique is confirmed as the training algorithm for the framework.

References

1. Thangavel SK, Bkaratki PD, Sankar A (2017) Student placement analyzer: a recommendation system using machine learning. In: 2017 4th International conference on advanced computing and communication systems (ICACCS). IEEE, pp. 1–5
2. Thabtah F, Abdelhamid N, Peebles D (2019) A machine learning autism classification based on logistic regression analysis. Health Inf Sci Syst 7(1):1–11
3. Nguyen C, Wang Y, Nguyen HN (2013) Random forest classifier combined with feature selection for breast cancer diagnosis and prognostic
4. Todorovski L, Džeroski S (2003) Combining classifiers with meta decision trees. Mach Learn 50(3):223–249

5. Palaniappan R, Sundaraj K, Sundaraj S (2014) A comparative study of the svm and k-nn machine learning algorithms for the diagnosis of respiratory pathologies using pulmonary acoustic signals. BMC Bioinf 15(1):1–8
6. Biau G, Cadre B, Rouvière L (2019) Accelerated gradient boosting. Mach Learn 108(6):971–992
7. Widodo A, Yang BS (2007) Support vector machine in machine condition monitoring and fault diagnosis. Mech Syst Signal Process 21(6):2560–2574
8. Rätsch G, Onoda T, Müller KR (2001) Soft margins for AdaBoost. Mach Learn 42(3):287–320
9. Chen T, Guestrin C (2016) Xgboost: a scalable tree boosting system. In: Proceedings of the 22nd ACM SIGKDD international conference on knowledge discovery and data mining, pp 785–794
10. Gavhane A, Kokkula G, Pandya I, Devadkar K (2018) Prediction of heart disease using machine learning. In: 2018 Second international conference on electronics, communication and aerospace technology (ICECA). IEEE, pp 1275–1278
11. Müller AC, Behnke S (2014) PyStruct: learning structured prediction in python. J Mach Learn Res 15(1):2055–2060
12. Huang Z (1997) Clustering large data sets with mixed numeric and categorical values. In: Proceedings of the 1st Pacific-Asia conference on knowledge discovery and data mining, (PAKDD), pp 21–34
13. Diamantidis NA, Karlis D, Giakoumakis EA (2000) Unsupervised stratification of cross-validation for accuracy estimation. Artif Intell 116(1–2):1–16
14. Islam MF, Ferdousi R, Rahman S, Bushra HY (2020) Likelihood prediction of diabetes at early stage using data mining techniques. In: Computer vision and machine intelligence in medical image analysis. Springer, Singapore, pp 113–125

Chapter 15
A Shallow Approach to Gradient Boosting ($XGBoost_s$) for Prediction of the Box Office Revenue of a Movie

Sujan Dutta and Kousik Dasgupta

Abstract In the recent past, machine learning paradigms like the ensemble approaches have been used effectively to predict revenue from large volumes of sales data that helped the decision-making process in many businesses. The proposed work in this paper proposes a modified approach of ensemble algorithms to predict box office revenues of upcoming movies. A shallow version of the gradient boosting ($XGBoost_s$) has been proposed to predict the box office revenue of movies based on several primary and derived features related to the movies in particular. Further studies have found that features such as budget, runtime, budget year ratio can also be considered as some of the more important estimators of the box office revenue. These features along with some other features have been used as an input to the proposed model in this proposed work to make significantly good predictions about the box office collection of a movie. The results are reported by testing and forecasting based on simulation on a standard data set. The precision of the model is tested using popular metrics such as R^2, MSLE. The results reported gives efficacy of the proposed approach that can be further used in other business models words.

Keywords Machine learning · Boosting · Box office revenue prediction · Supervised learning

1 Introduction

The movie industry is one of the biggest entertainment industries in the world. As of 2018, the global box office is worth $41.7 billion [1]. Every single day, a new movie is released somewhere around the world, with an expectation that it will be appreciated by the audience worldwide. And in today's age, movies are not just something people watch in their leisure time. Movies bring in necessary changes in the perspective and mentality of common people and thus, can't be completely avoided. But the cost of making a film is enormous. Thus, every single movie requires certain investors, in

S. Dutta (✉) · K. Dasgupta
Department of Computer Science and Engineering, Kalyani Government Engineering College, Kalyani, India

J. K. Mandal et al. (eds.), *Proceedings of International Conference on Innovations in Software Architecture and Computational Systems*, Studies in Autonomic, Data-driven and Industrial Computing, https://doi.org/10.1007/978-981-16-4301-9_16

the form of businessmen, etc. who is willing to invest his/her money into the making of the movie. But taking into consideration that 'Hollywood is the land of hunch and the wild guess' [2], it is not easy to get such investors because for investors, investing money into a film becomes a huge risk, which they are usually not ready to take. Jack Valenti, former president and CEO of the Motion Picture Association of America (MPAA), once said that "No one can tell you how a movie is going to do in the marketplace. Not until the film opens in a darkened theater and sparks fly up between the screen and the audience" [3]. Therefore, unless a movie hits the screens, it can never be told if it will be a successful movie or not, let alone the amount of revenue that the movie will be able to collect. But due to the massive growth of the film industry in recent times, there is a huge amount of data available on the internet that makes the revenue prediction somewhat possible.

In this research, we have collected one such dataset, containing the information about a movie that can be accessed during the production of the film, and using those values, we can somewhat accurately determine the amount of revenue that the movie can generate. We have used a modified Boosting Algorithm that uses the features of a movie to give the estimated revenue.

Boosting Algorithm seeks to improve the prediction power of a model by training a sequence of weak models (learners). Weak learners are trained sequentially and each learner tries to correct its predecessor. Notable variations of Boosting are listed as follows:

(a) AdaBoost (Adaptive Boosting): It combines many simple weak learners into one strong model [4]. The weak learners are decision stumps, i.e., decision trees having a single split. While creating the first learner, AdaBoost weighs all observations equally. After that, wrongly classified observations are assigned more weight than the correct ones to decrease the previous error. This process continues and it will keep adjusting the errors occurred in previous models until we find the best model.

(b) Gradient Boosting: Like AdaBoost, in Gradient Boosting [5] also we add simple predictors sequentially to an ensemble and each one tries to correct mistakes of the previous model. However, the Gradient Boosting method does not change the weights for wrongly classified observations at every iteration like Adaptive Boosting, instead it tries to fit a model to the residual errors produced by the previous model.

(c) XGBoost (eXtreme Gradient Boosting): XGBoost is an enhanced form of Gradient Boosting known for its speed and performance [6, 7]. XGBoost provides: (1) Parallelization of tree construction using all CPU cores, (2) Distributed Computation using a cluster of machines to train bulky models, (3) Out-of-Core Computing for large datasets, (4) Cache Optimization. XGBoost adds a custom regularization term in the objective function that is one of the main reasons for its effectiveness.

In this paper, we have used a modified approach of the XGBoosting algorithm for the prediction model.

The rest of the paper is organized as follows; this section has introduced the basics of the domain and prerequisites for training, validation, and testing of the proposed model. Section 2 describes previous work related to movie box office revenue prediction. Whereas Sect. 3 includes a description of dataset description and methodology and Sect. 4 explains model description and evaluation. Section 5 compares our model with some previous works. Finally, Sect. 6 concludes the article with a discussion of the efficacy of the proposed approach and future directions of research.

2 Literature Review

The popularity of a movie is an important factor in finding its revenue. One of the first attempts in trying to predict the popularity of a movie was made by Sreenivasan et al. [8]. It argued that movies with a higher level of novelty produce larger revenue. The work has suggested that there are many successful films with a high degree of novelty, which could be a manifestation of the inherent novelty preferences of the investors or risk-minimization based on some implicitly perceived inverted-U relationship between novelty and hedonic value.

Although Sreenivasan et al. were able to establish a relation between the degree of novelty and revenue, Sharda and Delen [9] were one of the first researchers to use Neural Network to predict box office collection of motion pictures and classify movies into nine categories according to their projected income, from "flop" to "blockbuster". Their results showed that the neural networks used in their study can predict the exact category of a movie before its theatrical release with 36.9% and within one category with 75.2% accuracy.

Later on, Lash and Zhao [10] proposed a Movie Investor Assurance System (MIAS) to provide predictions about the profitability of a movie. Their system automatically extracted important characteristics for a movie, including the cast and crew of the movie, topic of the movie, release date, etc. and then used those features to predict the success of the movie. According to them, the main contributions of their research was in 3 areas: (a) this decision support system was the first to harness machine learning, text mining, and social network analytics into one system to predict movie profitability during the early stages of movie production, and with minimal human interventions, (b) it proposed several novel features, e.g., dynamic network features, topic distributions, the match between "what" and "who", the match between "what" and "when", profit-based star power measures, etc. (c) their system was the first to collect different types of data (structured and unstructured data) from different publicly available sources, and combine them for predicting the success of a movie.

Asur and Huberman [11] introduced a very interesting and different approach in trying to predict how well a movie would do with the audience. In their paper, they showed how social media, especially Twitter, can be utilized to forecast future outcomes of movies. They were able to prove that with an increase in the number of tweets mentioning particular movies, it can be expected that the movie would also be appreciated by the audience, thus generating huge revenue. They constructed a linear regression model to predict the box office revenues of movies before their actual release [11].

A slightly different approach was adopted by Mestyán et al. [12] to predict the popularity of a movie. To estimate the popularity of a movie, they tracked down the movie's Wikipedia article (which was created before the actual release of the movie) and then used criteria like, 'number of views on the article', 'the number of human editors who have contributed to the article', 'number of edits made by human editors' (and many other similar features) to determine the popularity of the movie.

Hu [13] in his research has attempted to predict the domestic gross in movies with help of four (04) predictor variables viz. Budget, Distributor, Genre, and MPAA film rating system.

Eliashberg et al. [14] used movie scripts as input for their box office prediction model. The main contributions of this research were: (a) this paper was the first that analyzed actual movie scripts, (b) it showed that the kernel approach outperforms both regression and tree-based methods in the context of assessing box office performance, (c) the estimated "feature weights" provide insights about the important textual features for identifying useful "comps" for a new script.

In 2005, Delen and Sharda [15] implemented a decision support system for Hollywood managers called Forecast Guru. It employed different techniques including neural networks, decision trees, logistic regression, discriminant analysis, and information fusion algorithms to combine their results afterward.

Movie studios use tracking surveys [16] to estimate a movie's box office collection. Such surveys are telephonic interviews that track if public have heard of the movie and whether they plan to watch it. Based on these studies, movie studios have come up with their own prediction models. Generally, these models are not published. However, in 2014, the Sony Pictures tracking models got leaked in a hacker attack. These models were simple linear (mostly) combinations of the individual features coming from the tracking studies.

3 Dataset Description and Methodology

This section puts forward the methodology and details of the dataset used. The workflow diagram of the proposed approach is depicted in Fig. 1. The steps included in the proposed approach consist of data collection, data cleaning, feature transformation, feature extraction, and selection. The final stage is the generation of the Final proposed model.

Fig. 1 Workflow

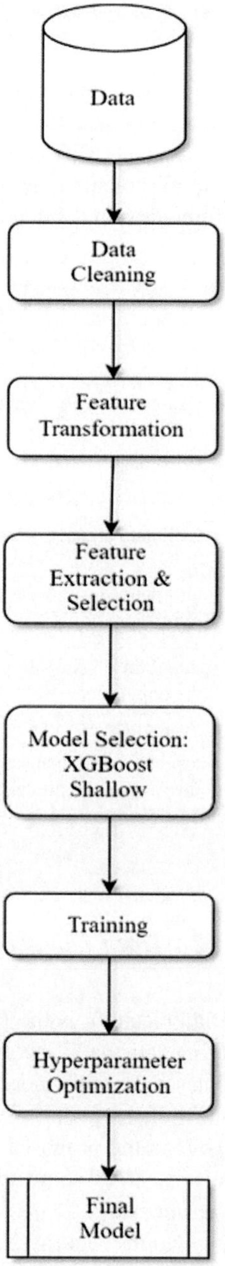

3.1 Data Collection

The dataset contains different features along with the revenues of 3000 movies. The data has been actually collected from the Movie DB API [17] which was provided on Kaggle [18]. There were some erroneous data which was later corrected with the help of the data retrieved from the Internet Movie Database (IMDb). The names of all features in the data set with its explanation are as given below:

• Belongs to collection—Name of the series	• Production countries—List of countries involved in the production of the movie
• Budget—Total budget	• Release date—Release date in the format "mm/dd/yy"
• Genres—List of suitable genres for the movie	• Runtime—Length of the movie
• Homepage—A link to the homepage	• Spoken languages—List of languages used
• IMDb id—Unique IMDb id for the movie	• Status—Released or not released
• Original language—Original language of the movie	• Tagline—Movie tagline
• Original title—Movie title	• Title—Movie title
• Overview—Overview of the movie	• Keywords—List of keywords
• Popularity—A floating-point value denoting the popularity	• Cast—List of actors
• Poster path—A link to the poster	• Crew—List of crew members
• Production companies—List of companies involved in the production of the movie	

3.2 Data Cleaning

Data cleaning is the process of preparing data for analysis by removing or modifying data that is incorrect, incomplete, irrelevant, or improperly formatted [19]. The dataset contains more than twenty (20) features corresponding to each movie but most of the features were in a format that is not suitable for feeding to the model. Thus in the proposed work, we had to clean some of them and discard the others. We discarded some unnecessary text-based data like *'id'*, *'plot'*, *'overview'*, *'crew'*, *'poster path'*, *'status'*, *'tagline'*, *'title'*, *'cast'*, *'production countries'*, *'production companies'*, etc., as the analysis of them gave reports that there is no significant correlation between these features and the revenue of the movie. Some of the features in the proposed work were reformatted properly like release date, presence of a home page, etc. to make it relevant in the proposed work. For example, the dates were converted into *'yyyy-mm-dd'* format, e.g., 05/20/99 became 1999-05-05. There were some year values like '2089' which clearly indicated an error from an empirical study that was corrected by introducing an offset of 100, thus '2089' became '1989'. We changed the binary feature (*'has homepage'*) that denotes the presence of a homepage into a binary number where, 0 means the movie doesn't have a homepage, 1 means the movie has a homepage.

3.3 Feature Transformation

Feature transformation is the process of replacing a feature with a function of itself for better learning [20]. On analysis, it was found that the data corresponding to 'budget' and 'revenue' is positively skewed so we applied $\ln(x + 1)$ transformation on both to make them manageable. Figures 2 and 3 shows the histogram plot of revenue data before transformations and after $\ln(x + 1)$ transformation respectively. Similarly,

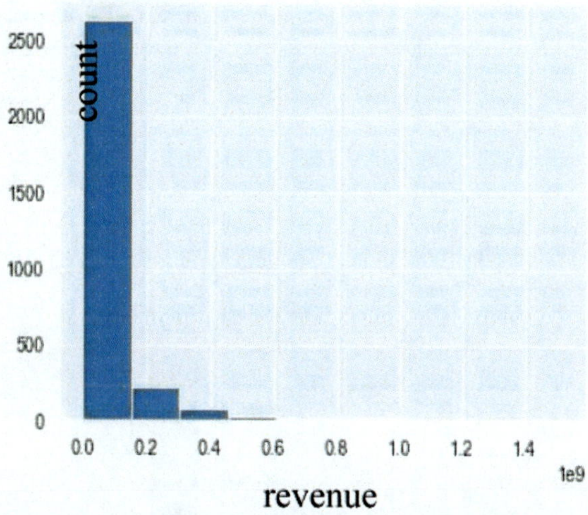

Fig. 2 Histogram plot of revenue feature showing skewed data

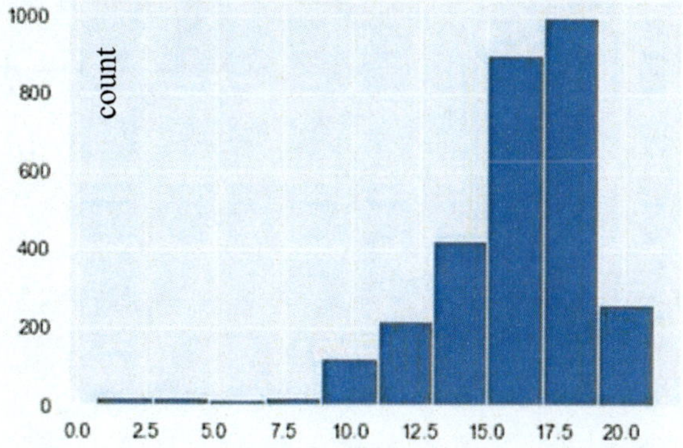

Fig. 3 Histogram plot of revenue feature after log transformation showing reduced skewness

and Figs. 4 and 5 show the budget feature after log [ln($x + 1$)] transformation. Thus the skewness is removed in both the features.

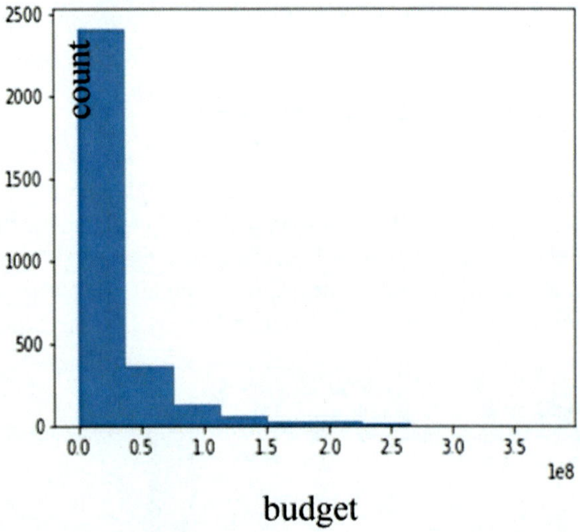

Fig. 4 Histogram plot of budge feature showing skewed data

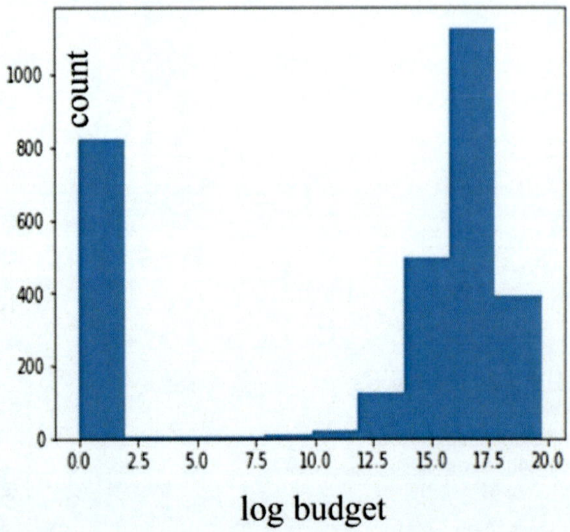

Fig. 5 Histogram plot of budget feature after log transformation showing reduced skewness

3.4 Feature Extraction and Selection

Feature selection is the process of selecting a subset of important features from the entire feature set. On the other hand, feature extraction is the process of extracting useful features from raw data [21]. In general, this step determines the features that may seem useless on the first look but can be useful after modification. Below is the description of how the features are modified (extracted) and selected.

From the *'date time'* we extracted the *'day of week'* and *'year'*. Then we applied one-hot encoding on the *'day of week'* feature. Figure 6 shows the distribution of film revenues over the days of the week. In this plot, the x-axis denotes the weekdays (numbered from 0–6) and the y-axis denotes the revenue ($\times 10^9$).

It was further found from Fig. 7, which depicts the histogram plot of top features that the ratio of *'budget'* to the square of *'release year'* acts as a good feature. It works because it provides a normalized budget. To find the most important features

Fig. 6 Scatter plot of revenue versus day of week

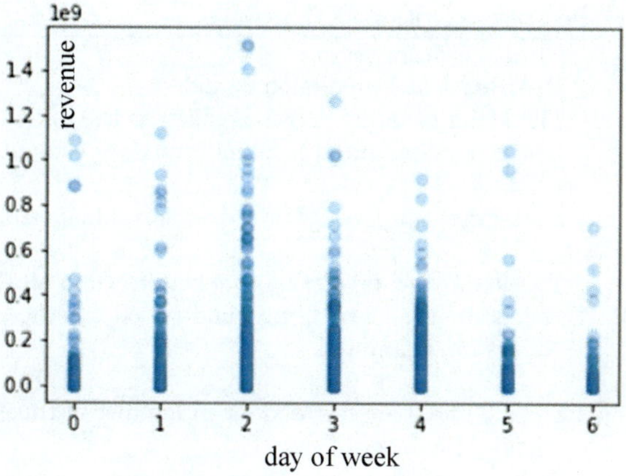

Fig. 7 Scatter plot of revenue versus day of week

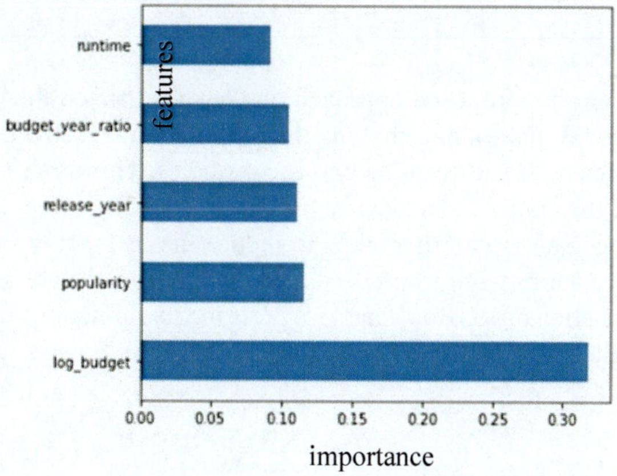

in the proposed work we have used an extra tree regressor [22]. Figure 7 shows the top five (05) important features.

Along with these five (05) features we used the one-hot encoded *'day of week'*, *'has homepage'*, and *'budget popularity ratio'* (ratio of budget to popularity) and *'year popularity ratio'* (ratio of release year to popularity). The last two features have been obtained in the same way as discussed above for the *'budget year ratio'* feature.

4 Model Description and Evaluation

The work in the paper proposes the use of a shallow version of a modified ensemble tree boosting algorithm named XGBoost [6]. The algorithm is given as below:

Algorithm 1: Proposed shallow version of modified ensemble tree boosting algorithm ($XGBoost_s$)

Input: Feature set
Output: Final prediction model

1. The Mean of target values is taken as the initial predictions and the corresponding initial residual errors are computed.
2. A new decision tree is trained with residual errors as the data.
3. To get the new prediction, predictions from all the previous trees are scaled by the learning rate and added together, and the new residual errors are computed.
4. Steps 2 and 3 are iterated for m number of times until we reach the required number of trees.
5. The final prediction is the additive sum of all previous predictions made by the trees.

In the proposed approach, we have restricted the depth of the decision trees to only 3 (three), thus making the proposed model shallow. The value of three (03) has been reported by doing empirical studies. This modification reduces the complexity of the model and makes it efficient. In step 2, the trees are built sequentially such that each subsequent tree aims to reduce the residual errors of the previous trees. Each tree learns from its predecessors and updates the residual errors. The error/objective function of XGBoost has two parts the loss function and the regularization term. The objective function at tth iteration is given in Eq. (1).

$$L^{(t)} = \sum_{t=1}^{n} l\left(y_i,\ \hat{y}_i^{(t-1)} + h_t(x_i)\right) + \Omega(h_t) \tag{1}$$

Table 1 Optimal hyperparameters

Hyperparameter	Optimal values
Gamma	0.1
Learning rate	0.095
Number of trees	200

Table 2 Performance evaluation with respect to MSLE and R^2

Name of set	Evaluation metrics	Value
Training	MSLE	3.02
	R^2	0.65
Validation	MSLE	3.53
	R^2	0.62
Test	MSLE	3.55
	R^2	0.63

where, $\hat{y}_i^{(t-1)}$ denotes the model we have got so far by ensembling $(t-1)$ trees. h_t represents the new tree. l is the loss function (in this work we have used root mean square error RMSE [23] as loss function). $\Omega(h_t)$ is the regularization term. y_i and x_i respectively denote the ground truths and input features.

To test the proposed model, the dataset has been split randomly into three sets; training, validation and test in the ratio 70:15:15. We used a grid search technique to find an optimal set of hyperparameters for the proposed model. Grid search has been used as an exhaustive search through the predetermined subset of hyperparameter space. The optimal values of hyperparameter for the proposed model are given in Table 1.

The proposed model has been evaluated using two metrics MSLE (Mean Squared Logarithmic Error) and R squared (R^2). The results obtained for both the metrics are given in Table 2. On comparing the performance of the model on training, validation, and test data, it can be said that the proposed model is effectively accurate and does not overfit the training data. The formula for MSLE is given in Eq. (2) and R^2 in Eq. (3) as below.

$$\text{MSLE} = \frac{1}{n}\sum_{i=1}^{n}(\log\ (y_i + 1) - \log(\hat{y}_i + 1))^2 \tag{2}$$

$$R^2 = 1 - \frac{\sum_{i=1}^{n}(y_i - \hat{y}_i)}{\sum_{i=1}^{n}(y_i - \overline{y})} \tag{3}$$

where, y_i is the ground truth and $\hat{y}_i$ is the predicted value. n is the total number of data points. $\bar{y}$ is the mean value of all y_i' s.

Table 3 Performance comparison

Model	MSLE
Linear regression [11]	5.50
Neural network (3 layer) [9]	4.31
Proposed model	3.55

5 Comparison to Other Models

The proposed work has been compared with other regression techniques in the literature. Specifically, we have compared with linear regression as used in [11] and a neural network approach as proposed in [9]. A linear regression model tries to fit a linear model to the observed data by minimizing the mean squared error between predictions and ground truths. A neural network is a multilayer perceptron model. Here, the error of the output layer is propagated backward and the weights are updated accordingly in each layer. We trained the linear regression model and the three-layer neural network on the same set of features mentioned earlier. The MSLE values of these models are compared with our model in Table 3. From the results of Table 3, it can be concluded that the proposed model outperforms both of them.

6 Conclusion and Future Works

In the proposed work a model has been proposed for the prediction of box office movie revenue collection. The approach can be used for forecasting of collection for upcoming movies. The features used for prediction are a mixture of derived and inherent features obtained from the movie that is to be predicted. We did not consider the crew (director, actors, etc.), the genre, story-line as features that may have some effect on the revenue. As future work, we are working on these features. However, the important features like *'budget'*, *'popularity'*, *'runtime'*, *'budget year ratio'* have been considered. We applied a modified approach of the XGBoosting algorithm for the prediction. Grid search technique was used to find an optimal set of hyperparameters for the proposed model. The final model is tested for its efficacy using standard performance measures R^2, MSLE. Also, the proposed system is compared with a version of linear regression and neural network. The results obtained are encouraging.

However, sometimes the revenue generated by a movie does not only depend on the parameters that we have considered. There are some external conditions that affect revenue considerably. For example, the socio-economic condition of a country affects the box office collection. Another factor that may affect revenue is the piracy of movies. Both the conditions actually affect the total count of the audience actually spending their money at the theaters; thus, the box office collection will be affected. The future work in the proposed system can critically use these futures for more accurate predictions.

References

1. Variety.com (2018) Worldwide box office hits record as Disney dominates. https://variety.com/2019/film/news/box-office-record-disney-dominates-1203098075. Last accessed 05 Nov 2020
2. Litman BR (1998) The motion picture mega-industry. Allyn & Bacon
3. Valenti J (1978) Motion pictures and their impact on society in the year 2001. Midwest Research Institute
4. Freund Y, Schapire RE (1997) A decision-theoretic generalization of on-line learning and an application to boosting. J Comput Syst Sci 55(1):119–139
5. Friedman JH (2001) Greedy function approximation: a gradient boosting machine. Ann Stat 1189–1232
6. Chen T, Guestrin C (2016) Xgboost: a scalable tree boosting system. In: Proceedings of the 22nd ACM SIGKDD international conference on knowledge discovery and data mining, pp 785–794
7. GitHub. https://github.com/dmlc/xgboost. Last accessed 05 Nov 2020
8. Sreenivasan S (2013) Quantitative analysis of the evolution of novelty in cinema through crowdsourced keywords. Sci Rep 3(1):1–11
9. Sharda R, Delen D (2006) Predicting box-office success of motion pictures with neural networks. Expert Syst Appl 30(2):243–254
10. Lash MT, Zhao K (2016) Early predictions of movie success: the who, what, and when of profitability. J Manag Inf Syst 33(3):874–903
11. Asur S, Huberman BA (2010) Predicting the future with social media. In: 2010 IEEE/WIC/ACM international conference on web intelligence and intelligent agent technology, vol 1. IEEE, pp 492–499
12. Mestyán M, Yasseri T, Kertész J (2013). Early prediction of movie box office success based on Wikipedia activity big data. PloS ONE, 8(8):e71226
13. Berkely Edu, Domestic gross of movies. https://www.stat.berkeley.edu/~aldous/Research/Ugrad/Xiaoyu_Hu.pdf. Last accessed 05 Nov 2020
14. Eliashberg J, Hui SK, Zhang ZJ (2014) Assessing box office performance using movie scripts: a kernel-based approach. IEEE Trans Knowl Data Eng 26(11):2639–2648
15. Delen D, Sharda R, Kumar P (2007) Movie forecast Guru: a web-based DSS for Hollywood managers. Decis Support Syst 43(4):1151–1170
16. Pope LS, Jason E (eds) (2017) The movie business book. Routledge (A Focal Press Book), New York, pp. xxiii, 628. ISBN 978-1-138-65629-1
17. The Movie Database API. https://developers.themoviedb.org. Last accessed 05 Nov 2020
18. Kaggle TMDB box office prediction. https://www.kaggle.com/c/tmdb-box-office-prediction/data. Last accessed 05 Nov 2020
19. Rahm E, Do HH (2000) Data cleaning: problems and current approaches. IEEE Data Eng Bull 23(4):3–13
20. EDSA The Essentials of Data Analytics and Machine Learning. https://courses.edsa-project.eu/pluginfile.php/1332/mod_resource/content/0/Module%205%20-%20Feature%20transformation_V1.pdf. Last accessed 05 Nov 2020
21. Guyon I, Gunn S, Nikravesh M, Zadeh LA (eds) (2008) Feature extraction: foundations and applications, vol 207. Springer
22. Geurts P, Ernst D, Wehenkel L (2006) Extremely randomized trees. Mach Learn 63(1):3–42
23. Chai T, Draxler RR (2014) Root mean square error (RMSE) or mean absolute error (MAE)?–arguments against avoiding RMSE in the literature. Geosci Model Dev 7(3):1247–1250

Chapter 16
A Deep Learning Framework to Forecast Stock Trends Based on Black Swan Events

Samit Bhanja and Abhishek Das

Abstract The stock trends prediction is the key interest area for the investors. If the successful stock trends prediction is achieved, then the investors can adopt a more appropriate trading strategy, and that can significantly reduce the risk of investment. But it is hard to predict the stock market due to the unpredictable fatal events called Black Swan events. In this work, we propose a deep learning framework to predict the daily stock market trends with the intent that our framework can predict the stock market even on the time periods of the Black Swan events. In this framework, the signals of various technical indicators along with the daily closing price of the stock market and other influencing stock markets are used as the input for more accurate predictions. The base module of this framework is 1D convolutional neural network (1D-CNN) and bidirectional gated recurrent unit (Bi-GRU) neural network. We conduct vast experiments on the real-world datasets from two different stock markets and show that our framework exhibit satisfactory prediction accuracy for the normal circumstances. It outperforms other existing similar works during the periods of Black Swan events.

Keywords Stock trends prediction · Deep learning · 1D convolutional neural network · Bidirectional gated recurrent unit · Black Swan event

1 Introduction

Prediction is one of the biggest challenges for any research area. If a successful prediction is achieved, then that can smoothen the livelihood of the people. One of the most challenging time-series prediction research area is the stock trend prediction because of its nonlinearity and highly dynamic behavior. The stock market is also

S. Bhanja (✉)
Government General Degree College, Singur, Hooghly 712409, India

A. Das
Aliah University, New Town, Kolkata 700160, India
e-mail: adas@aliah.ac.in

J. K. Mandal et al. (eds.), *Proceedings of International Conference on Innovations in Software Architecture and Computational Systems*, Studies in Autonomic, Data-driven and Industrial Computing, https://doi.org/10.1007/978-981-16-4301-9_17

suffered by unpredictable events that have a serious negative impact on it are called the Black Swan events. If the successful stock trend is predicted, then the stock market regulators and the investors can develop better trading strategies, which helps them to reduce the risk of the investment and also increase profitability.

In recent times, numerous research work has been done to develop the most appropriate model for the stock market prediction. These stock market predictions have been done through two types of analysis, viz. fundamental analysis and the technical analysis. The fundamental analysis [1, 2] uses various types of economic and financial information, such as macroeconomic factors, foreign exchange rates, etc. On the other hand, technical analysis [3, 4] is done based on the stock price information. Currently, the globalization effect increases the financial flow across the world, and as a result, the stock market of a country is influenced by the other international stock market. There are a limited number of efforts [5, 6] have been done in this regard. But they are not considered the different international Black Swan events that highly influence the stock market. Nowadays, deep learning methods have become the most popular approach for the stock market prediction. These deep learning approaches outperform traditional machine learning methods [7]. Deep recurrent neural networks (RNNs) are one of the most suitable ANNs for the time-series classification and regression problems. But still, they are suffering from the vanishing and exploding gradient problems [8]. Long short-term memory (LSTM) and gated recurrent unit (GRU) neural networks are two variations of RNN that overcome these problems of the traditional RNNs. Both LSTM and GRU are competent to extract the complex temporal features from the long-sequence time-series data. On the other hand, CNNs are excellent for extracting the spatial features from an input window of the time-series dataset [9].

In this study, we propose a unique deep learning-based stock trend prediction framework (DSTPF). This framework addresses the correlation among other international stock markets and also addresses the different Black Swan events of the stock market. Here we also consider the different technical indicators of the stock market. As part of this framework, we have built a hybrid deep learning models by combining 1D convolutional neural network (1D-CNN) and bidirectional gated recurrent unit (Bi-GRU) architectures to extract the spatial and temporal features from the nonlinear multivariate stock market time-series data. The purpose of this research work is to build a powerful deep learning framework that can predict the stock trends with a high accuracy intensity, even on the time-period of Black Swan events. This research work will able to help the investors and the regulator of the financial stock market.

We organize the rest of the paper as follows: In Sect. 2, we represent the concise description of the related work. The detailed description of our proposed framework is presented in Sect. 3. In Sect. 4, we describe the experimental process and compare our framework with other state-of-the-art models. Finally, in Sect. 5, the conclusion is drawn.

2 Literature Review

In the last few decades, many efforts were made to predict stock market trends. These efforts are mainly classified into fundamental analysis and technical analysis. The fundamental analysis uses different macro-economic factors for the prediction, on the other hand, the technical analysis method [10] uses the historical stock price for the prediction. Later on, these technical analysis methods have been formulated as a pattern recognition problems. These pattern recognition problems to predict the stock market are broadly divided into statistical methods, machine learning methods, deep learning methods, and hybrid methods [11–13] that combine two or more different types of methods.

Autoregressive integrated moving average (ARIMA) model is a popular machine learning for forecasting of linear time-series data. In [14], ARIMA model was used for short-term stock trends prediction. In [15], the authors proposed the ARIMA-BP model by combining the ARIMA with the back propagation (BP) neural network model to capture both the linearity and nonlinearity of the stock market time-series data for the stock trends prediction. Here, the authors showed that this ARIMA-BP model generates better prediction accuracy compared to the shallow ARIMA and BP model.

In recent times, with the increased computational power and the availability of large volumes of historical data, deep learning architectures become one of the popular tool for the time-series prediction. A large number of deep learning-based novel research works have been done to predict the stock market trends. In [16], a deep learning model was proposed by combining the CNN and LSTM architectures. Here, the authors used the technical indicators and the textual information for the stock market prediction to improve the prediction performance. The LSTM model was applied to the large volume of the S&P500 dataset in [17], and it proved that the LSTM model is superior compared to the standard deep neural network architectures. In [7], authors integrated 1D-CNN with LSTM architecture to build a deep learning model and showed that the combined architecture is better than the shallow architectures. The LSTM, RNN, and GRU architectures were compared in [18] to forecast the Google stock price movement. Here, the authors showed that LSTM and GRU are exhibited better prediction accuracy compared to RNN. The authors in [19] proved that the combined model of Wavelets and CNN is better than the traditional ANN approaches. The multiple pipeline model was developed in [20] by combining CNN and bidirectional LSTM. Here, it has been established that the multiple pipeline model's performances were increased by 9% over a single pipeline model. In [21], the authors proposed a motifs extraction algorithm to extract the higher-order structural features from the raw time-series data. These motifs were applied to the CNN model to extract the abstract features to predict stock trends. The authors also showed that the model improves the prediction accuracy compared to the other traditional methods.

However, most of the recent work considered the stock market prediction as a pattern recognition problem, and to improve the prediction accuracy, they concentrated on the selection of the different types of machine learning, deep learning, and/or statistical models. Some of the research work was concentrated on the analysis of the various technical indicators and sentiments of the tweeter and news data which are related to the stock market. Differentiating from the existing work, here we proposed a unique global event-based deep learning framework that finds how a stock market is affected by the other international stock markets based on the various unpredictable global events. And from this local and global stock market time-series data, we tried to find out the spatial and temporal features by a hybrid deep learning model to predict the stock market trends.

3 Proposed Framework

In this section, we describe in detail our proposed deep learning framework DSTPF. The flowchart of our proposed framework is shown in Fig. 1. The overall work is mainly divided into five steps.

3.1 Data Description and Correlation Analysis

Stock market forecasting is crucial to take the early step to control the market. Here, to predict a stock market, we collect the historical data of that stock market, which is considered as the local stock market. Due to the globalization effect, the stock index of a market is highly dependent on the other global stock markets. So, in this work, we also collect the historical stock market data of other global stock markets. Here to analyze the effectiveness of each global stock market on the local market, we find the correlation between them. Here to identify the correlation, we calculate the Pearson correlation coefficient by the following formula.

$$R = \frac{m \sum v_1 v_2 - (\sum v_1)(\sum v_2)}{\sqrt{[m \sum v_1^2 - (\sum v_1)^2][m \sum v_2^2 - (\sum v_2)^2]}} \tag{1}$$

Here v_1 and v_2 are two vectors, and m is number of information. The global stock markets which are highly correlated to the local stock market are considered for the local stock market trends prediction.

3.2 Black Swan Event Analysis

In the world economy, the unpredictable events that have a severe negative impact on the global economy and which are extremely rare are called the Black Swan event.

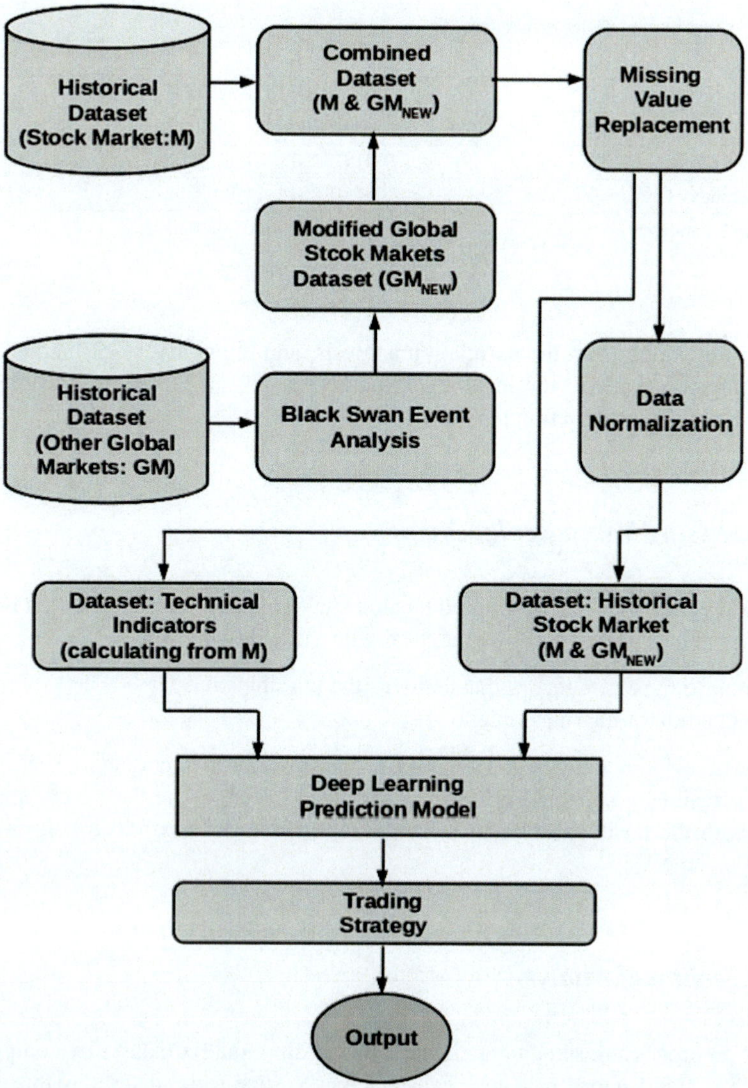

Fig. 1 Proposed framework DSTPF

In general, the standard forecasting tools are not capable to predict the destructibility of these events.

In this study, to develop a robust forecasting model, we analyze the effect of various Black Swan events on the local stock market. In this regard, we consider the five devastating global Black Swan events as presented in Table 1.

Here, we analyze how the global stock markets (selected in Sect. 3.1) are reacting to the local stock market with the time periods of different Black Swan events. In

Table 1 Description of Black Swan events

Event	Duration
Dot-Com Crash	01.03.2001–31.12.2004
Global Financial Meltdown	01.09.2008–31.12.2009
Crude Oil Crisis	01.01.2014–31.12.2014
Black Monday China	01.01.2014–31.12.2014
COVID19 Pandemic	01.12.2019–30.04.2020

this regard, we perform the correlation analysis, and the global stock markets that are highly correlated to the local stock market for all the Black Swan events are considered for the prediction purpose.

3.3 Missing Value Replacement

The most common problem suffered by almost all domains of time-series data is the missing value. The two ways we can deal with these missing values

1. Remove the entire record that contains the missing value.
2. Imputation of missing values.

For any time-series analysis, the temporal properties play an important role, so option 1 is not feasible. So, in this work, we adopt option 2. Here, we replace every missing value with the mean value of its immediate previous and next day's non-missing values.

3.4 Technical Analysis

In this preprocessing step, we generate a new feature matrix dataset by computing the technical indicators from the local stock market time-series dataset. In this work, we use four different technical indicators [22, 23]. The formula of the technical indicators is depicted in Table 2.

3.5 Prediction Model

Figure 2 represents the deep learning-based prediction model (DPM) that we construct for our proposed framework. The reason behind the selection of deep learning compared to machine learning is that deep learning has automatic feature extraction

Table 2 Formula of technical indicators

Technical indicator	Formula	Technical indicator	Formula
$MAI(n1, n2)$	$\frac{SMA(n1)}{SMA(n2)}$	$\%D$	$(D-1) \times \frac{2}{3} + \%K \times \frac{1}{3}$
$RSV(n)$	$\frac{C-Low_n}{High_n-Low_n} \times 100$	$RSI(n)$	$\frac{U}{U+D} \times 100$
$\%K$	$(K-1) \times \frac{2}{3} + RSV \times \frac{1}{3}$	$ROC(n)$	$\frac{C-C_n}{C_n} \times 100$

–$SMA(n1)$and $SMA(n2)$ are respectively the $n1$ and $n2$ days moving average. Here, $n1 < n2$

–C is the current-day closing price, Low_n and $High_n$ are respectively the n days low price and high price

–$(K-1)$ and $(D-1)$ are respectively the K value and D value of the previous day

–U and D respectively denotes the average of n days up and down of the closing price

–C_n represents the closing price of n days before

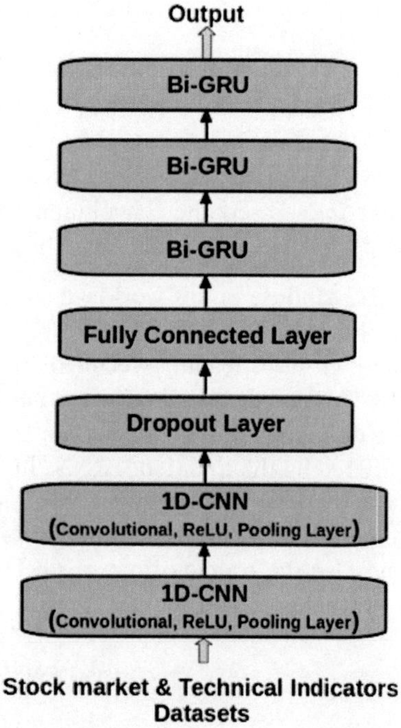

Fig. 2 Deep learning-based prediction model (DPM)

capabilities from the large volume of the raw historical dataset, whereas machine learning needs the help of data scientists for the handcrafted features extraction. In this prediction model, we used both the raw stock market dataset and stock market features (technical indicators) dataset as the input for the intention of more accurate prediction. DPM is constructed by combining the two most popular deep learning architectures, viz. 1D convolutional neural network (1D-CNN) and bidirectional gated recurrent unit (Bi-GRU). This model consists of two modules, viz. CNN module and GRU module.

CNN Module The CNN module is constructed by two 1D-CNN layers, a dropout layer and a fully connected layer, whereas each 1D-CNN layer has three sublayers, viz. convolutional, activation (ReLU), and pooling (max-pooling) sublayers. The input of the first 1D-CNN layer is the raw stock market data and the technical indicators of the stock market. The output of the first 1D-CNN layer is compressed local features which are taken as the input of the next 1D-CNN layer. The output of the second 1D-CNN layer is a higher-level feature matrix. Then the dropout layer randomly removes some of the model parameters to reduce the possibilities of overfitting. In the next fully connected layer, we flatten the feature matrix.

Here the computational formula for the 1D-CNN layer is as follows:

$$O_i^l = \sum X_i^{l-1} \otimes F_{i,j} + B_j^l \tag{2}$$

$$Y_i^l = ReLU(O_i^l) \tag{3}$$

$$X_i^l = \text{Pool}(Y_i^l) \tag{4}$$

Here $\otimes$ represent the convolution operation, $F_{i,j}$ denotes the filter. X, B, and l are respectively denoting the input, bias, and the layer number.

GRU Module In this work, we use the three layers of bidirectional GRU(Bi-GRU) as shown in Fig. 2. The GRU module is used to extract the temporal features from the high-level features vector of the CNN module. The long short-term memory (LSTM) and the gated recurrent unit (GRU) [24], two variants of traditional RNNs, are developed to solve two major problems, viz. vanishing and exploding gradients suffered by the traditional RNN. In this study, we select GRU over LSTM because GRU has fewer gates than LSTM, so it required lass computation. Moreover, the research [25] shows that the performance of GRU is better compared to LSTM, and also the training time of GRU is less. GRU uses the following formula for computational purposes.

$$ug_t = \delta(V_{ug} * [x(t), p(t-1)]) \tag{5}$$

$$rg_t = \delta(V_{rg} * [x(t), p(t-1)]) \tag{6}$$

$$\hat{p}(t) = \delta(V_p * [x(t), (rg_t * p(t-1))]) \tag{7}$$

$$p(t) = (1 - ug_t) * p(t-1) + ug_t * \hat{p}(t) \tag{8}$$

Here, ug_t and rg_t are respectively the output of update gate and reset gate. $x(t)$ and $p(t-1)$ are respectively the input at the current time stamp and output of the previous time stamp, and δ is the activation function. The weights of the update gate, reset gate, and the contender output are V_{ug}, V_{rg}, and V_p respectively.

In this work, despite traditional GRU, we use the bidirectional GRU (Bi-GRU) which is constructed by two traditional GRUs, where one GRU processes the time-series data in the forward direction and the other GRU in the backward direction, and then integrates them to produce the final output. Thus, the Bi-GRU learns the future features from the previously observed data and the past features by observing the future data. In this way, the Bi-GRU extracts both ways temporal features from the higher-order feature vectors stock market data.

3.6 Trading Strategy

In this section, we adopt a trading strategy to help the investors and the regulators of the stock market. In this trading strategy, we generate three trading rules—buying, selling, and retaining. Here the output of the prediction model is analyzed to generate the trading signal by using the following formula.

$$\text{Incr}_{t+1} = \frac{(\text{pred}_{t+1} - \text{act}_t)}{\text{act}_t} \times 100\% \tag{9}$$

$$T\_\text{Sig}_{t+1} = \begin{cases} 1, & \text{if} \text{Incr}_{t+1} \geq \theta \\ -1, & \text{if} \text{Incr}_{t+1} \leq -\theta \\ 0, & \text{Otherwise} \end{cases} \tag{10}$$

where act_t is the actual value at time stamp t, prd_{t+1} is the predicted value at the next time stamp, θ is the threshold value(%), and T_Sig_{t+1} is the trading signal. Here, we represent 1 as buying, -1 as selling, and 0 as retaining signal.

4 Experiment

In this section, the experimental process is described in detail. We use two real-world datasets for the experimental process of our proposed framework DSTPF. Here, we also compare our proposed DSTPF with the other state-of-the-art models for analyzing and validating of our proposed work.

4.1 Dataset Description

A detailed description of the datasets which are used for the simulation of our research work is presented in Table 3. All these stock market datasets contain various attributes, viz. opening price, high price, low price, closing price, volume, and adjusted closing price. Here we only consider the closing price of the stock market.

Table 3 Dataset description

Local StockMarket	Global StockMarkets	Time period	Source
Bombay stock exchange (BSE)	S&P500, HSI, NIKKEI225, KOSPI, FTSE100, SSE, DJI, NASDAQ, NYSE	01.01.2000–30.04.2020	Yahoo Finance [26]
US stock market S&P500	BSE, HSI, NIKKEI225,KOSPI, FTSE100, SSE, DJI, NASDAQ, and NYSE		

4.2 Experimental Setup

In this work, we conduct all the experiments in Python programming language with the support of Keras open-source deep-learning library.

In this research work, we use the previous 7 days' stock market data to predict the 8th day's closing price. Here five technical indicators such as MAI(10,20), MAI(10,40), KD, RSI(20), and ROC(40) are determined form the forecasting dataset as per the formula in Table 2. The raw stock market dataset is normalized by $Tanh\ Estimators$ [27] normalization method to scale down the data to $[1, -1]$ range.

The splitting of the dataset into train–test plays a vital role for any machine learning approach. We need to split the dataset in such a way that both the training and testing dataset must contain the usual and unusual instances of the problem domain. By considering this criterion, we divide the dataset by the $train\_test\_split()$ function, where 70% of data used for the training purpose and the remaining 30% of data used for the testing purpose. Here we also separately test the models on the Crude Oil Crisis, Black Monday China, and COVID19 Pandemic events' time-periods data. In this experiment, the stock market trends are predicted for three different threshold values ($\theta = 1\%$, $\theta = 2\%$, and $\theta = 3\%$). To get the best prediction result from our prediction model DPM, we fine-tuned our model by varying the deep learning parameters. After the fine-tuning process, we set the model parameters, which are presented in Table 4.

4.3 Evaluation

In this work, we adopt three performance evaluation matrices, viz. root mean square error ($RMSE$), mean absolute error (MAE), and $Accuracy(\%)$. $RMSE$ and MAE are used to analyze the performance of the forecasting models, whereas $Accuracy$ is used to evaluate the performance of the proposed forecasting framework DSTPF. These matrices are calculated as per the formula below

Table 4 Prediction model's(DPM) Parameters.

Module	Layer	Parameters
CNN module	1st 1D-CNN layer	Filters:80, Kernel size: 8, stride: 1
	2nd 1D-CNN layer	Filters:64, Kernel size: 4, stride: 1
	Droupout layer	dropout rate:0.5
	Fully connected layer	Neuron: 64
GRU module	1st Bi-GRU layer	Neurons: 64
	2nd Bi-GRU layer	Neurons: 32
	3rd Bi-GRU layer	Neurons: 16
Training batch size: 16, no. of epochs: 100, optimizer: ADAM		

$$\text{RMSE} = \sqrt{\frac{1}{n}\sum_{k=1}^{n}(\text{acc}_k - \text{pred}_k)^2} \tag{11}$$

$$\text{MAE} = \frac{1}{n}\sum_{k=1}^{n}|\text{acc}_k - \text{pred}_k| \tag{12}$$

$$\text{Accuracy}(\%) = \frac{\text{no. of true predictions}}{\text{total no. of predictions}} \times 100\% \tag{13}$$

where, acc_k and pred_k are the kth actual value and predicted value respectively.

In this work, to analyze the effectiveness of our proposed forecasting model (DPM), we compare its performance with the shallow SVR, ARIMA, LSTM, and GRU models for same input dataset with same data preprocessing environment. Here to evaluate the performance of our framework, we compare it with other popular models such as ARIMA-BP [15], DISM [7], CNN [21], and LSTM [17]. The parameters that are used for each of the models are presented in Table 5.

4.4 Results and Discussion

In Table 6, we compare our proposed forecasting models with the baseline models. From this table, we can see that our forecasting model (DPM) has the lowest prediction errors. It is also observed that deep learning models are better compared to the traditional time-series analysis models (SVR & ARIMA). The combination of CNN and GRU models can extract both the abstract and temporal features from the time-series data and due to this reason, they are more convenient to handle the time-series data (Fig. 3).

Table 5 Parameter list of different models

Model name	Parameters
SVR	kernel = rbf, C = 50, gamma = 0.1
ARIMA	$p = 7, d = 1, q = 1$
Shallow LSTM	neurons = 80, activation = sigmoid, dropout = 0.3,batch size = 16, epoch = 150
Shallow GRU	neurons = 72, activation = sigmoid, dropout = 0.3,batch size = 20, epoch = 100
ARIMA-BP [15]	ARIMA: $p = 7, d = 2, q = 1$ BP: (16, 64, 1) neurons
DISM [7]	1D-CNN layer = 2, LSTM layer = 3, batch size = 16, dropout = 0.4, epoch = 100
LSTM [17]	LSTM layer = 3, Batch size = 20, dropout = 0.3, epoch = 100
CNN [21]	1D-CNN layer = 3, Dense layer = 1, dropout = 0.3, batch size = 16

Table 6 Comparisons of testing errors of different forecasting models

Forecasting model	BSE index		S&P500 index	
	RMSE	MAE	RMSE	MAE
SVR	0.068235	0.052913	0.076214	0.05421
ARIMA	0.067507	0.052793	0.074934	0.052982
Shallow LSTM	0.061310	0.046211	0.071028	0.051792
Shallow GRU	0.060810	0.045019	0.070883	0.051226
DPM (our)	**0.044113**	**0.029102**	**0.056851**	**0.035943**

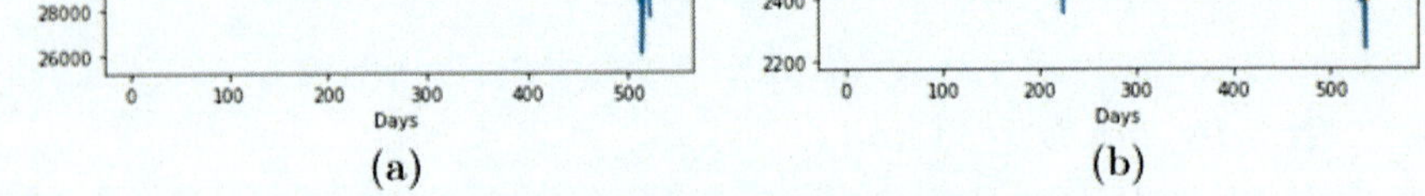

Fig. 3 Forecasted results of our proposed framework DSTPF **a** BSE and **b** S&P500

In Fig. 4, we graphically compare the stock trends prediction accuracy of our proposed framework DSTPF with other well-established models for various thresholds. From both figures, we can observe that our proposed framework increased the prediction accuracy by 8–12% compare to ARIMA-BP, DISM, and LSTM mod-

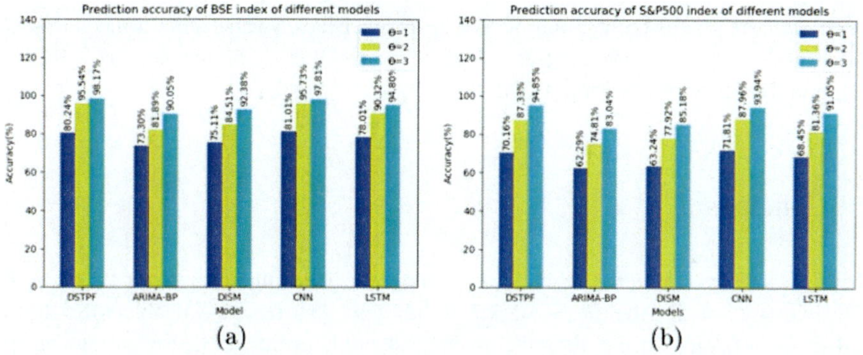

Fig. 4 Stock trends prediction accuracy comparison. **a** BSE and **b** S&P500

Table 7 Accuracy (%) comparison of various models to forecast BSE stock trends analogous to various Black Swan events for various thresholds (θ)

Model	Oil crisis			Black Monday			COVID19 pandemic		
	acc(%) ($\theta = 1$)	acc(%) ($\theta = 2$)	acc(%) ($\theta = 3$)	acc(%) ($\theta = 1$)	acc(%) ($\theta = 2$)	acc(%) ($\theta = 3$)	acc(%) ($\theta = 1$)	acc(%) ($\theta = 2$)	acc(%) ($\theta = 3$)
DSTPF	**93.23**	**98.96**	**99.84**	**91.79**	**96.57**	**99.28**	**97.21**	**98.01**	**98.49**
ARIMA-BP	69.41	80.79	83.27	70.52	83.46	87.21	72.74	83.40	86.91
DISM	72.05	82.32	85.91	71.53	83.97	89.34	73.01	84.82	88.93
CNN	91.51	96.84	98.29	90.73	94.18	97.38	95.37	96.19	97.22
LSTM	78.51	88.03	95.71	83.67	89.14	95.30	87.11	90.28	93.61

The bold face information depicts the results of the proposed method DSTPF in comparison with existing methods

Table 8 Accuracy (%) comparison of various models to forecast S&P500 stock trends analogous to various Black Swan events for various thresholds (θ)

Model	Oil crisis			Black Monday			COVID19 pandemic		
	acc(%) ($\theta = 1$)	acc(%) ($\theta = 2$)	acc(%) ($\theta = 3$)	acc(%) ($\theta = 1$)	acc(%) ($\theta = 2$)	acc(%) ($\theta = 3$)	acc(%) ($\theta = 1$)	acc(%) ($\theta = 2$)	acc(%) ($\theta = 3$)
DSTPF	**95.20**	**98.92**	**99.89**	**93.13**	**98.84**	**99.75**	**96.78**	**97.60**	**97.93**
ARIMA-BP	71.37	82.09	86.26	70.85	77.55	84.07	77.18	83.48	86.19
DISM	75.83	83.97	90.10	72.00	79.53	87.37	76.29	83.60	87.15
CNN	93.51	97.11	98.42	91.99	96.97	98.32	94.98	96.34	95.87
LSTM	89.31	92.05	93.33	87.19	89.64	92.53	88.00	91.08	92.83

The bold face information depicts the results of the proposed method DSTPF in comparison with existing methods

els, whereas the accuracy of our framework is almost the same as the CNN model. In Tables 7 and 8, we compare the prediction accuracy of our framework with the other models analogous to various Black Swan events to forecast BSE and S&P500 stock trends, respectively. Our framework explodes the interrelation among various Black Swan events with the stock market. Due to this fact, our framework produced more accurate results corresponding to the various Black Swan events' time periods. We can also observe that our framework increases the prediction accuracy 1–2% compared to its archrival model CNN.

5 Conclusion

In this work, we intended to build a robust deep learning framework that could efficaciously predict the stock market trends and also have been able to exhibit stable performance during unpredictable catastrophic events. In this paper, we have proposed a deep learning framework DSTPF. Here, we analyze the various Black Swan events to identify underlying relationships with the global stock markets.

In comparison with other well-established models, our framework has exhibited stable performance in terms of prediction accuracy and outperforms other models for the Black Swan events' time periods. From these experiments, we have concluded that our framework not only predicts the stock trends with a high degree of accuracy in normal circumstances but also for unpredictable catastrophic situations. In this work, the proposed framework is only tested for predicting the stock index of BSE and S&P500 stock markets. In the future, we like to predict the individual stock price for different companies and also like to improve the prediction accuracy by adding other relevant factors.

References

1. Dechow PM, Hutton AP, Meulbroek L, Sloan RG (2001) Short-sellers, fundamental analysis, and stock returns. J Financ Econ 61(1):77–106
2. Shen KY, Tzeng GH (2015) Combined soft computing model for value stock selection based on fundamental analysis. Appl Soft Comput 37:142–155
3. Mizuno H, Kosaka M, Yajima H, Komoda N (1998) Application of neural network to technical analysis of stock market prediction. Stud Inf Control 7(3):111–120
4. Chenoweth T, ObradoviĈ Z, Lee SS (2017) Embedding technical analysis into neural network based trading systems. In: Artificial intelligence applications on wall street. Routledge, pp 523–541
5. Jiang X, Pan S, Jiang J, Long G (2018) Cross-domain deep learning approach for multiple financial market prediction. In: 2018 international joint conference on neural networks (IJCNN), pp 1–8
6. Mukherjee D (2007) Comparative analysis of Indian stock market with international markets. Great Lakes Herald 1(1):39–71
7. Bhanja S, Das A (2019) Deep learning-based integrated stacked model for the stock market prediction. Int J Eng Adv Technol 9(1):5167–5174
8. Bengio Y, Simard P, Frasconi P et al (1994) Learning long-term dependencies with gradient descent is difficult. IEEE Trans Neural Netw 5(2):157–166
9. Wang Z, Yan W, Oates T (2017) Time series classification from scratch with deep neural networks: a strong baseline. In: 2017 international joint conference on neural networks (IJCNN), pp 1578–1585
10. Murphy J (1999) Technical analysis on the financial markets. New York institute of finance

11. AbdelKawy R, Abdelmoez WM, Shoukry A (2021) A synchronous deep reinforcement learning model for automated multi-stock trading. Progr Artif Intell, 1–15
12. Yu P, Yan X (2020) Stock price prediction based on deep neural networks. Neural Comput Appl 32(6):1609–1628
13. Yuan X, Yuan J, Jiang T, Ain QU (2020) Integrated long-term stock selection models based on feature selection and machine learning algorithms for china stock market. IEEE Access 8:22672–22685
14. Kamalakannan J, Sengupta I, Chaudhury S (2018) Stock market prediction using time series analysis. Comput Commun Data Eng Ser 1(3)
15. Du Y (2018) Application and analysis of forecasting stock price index based on combination of Arima model and BP neural network. In: 2018 Chinese control and decision conference (CCDC), pp 2854–2857
16. Oncharoen P, Vateekul P (2018) Deep learning for stock market prediction using event embedding and technical indicators. In: 2018 5th international conference on advanced informatics: concept theory and applications (ICAICTA), pp 19–24
17. Fischer T, Krauss C (2018) Deep learning with long short-term memory networks for financial market predictions. Eur J Oper Res 270(2):654–669
18. Di Persio L, Honchar O (2017) Recurrent neural networks approach to the financial forecast of google assets. Int J Math Comput Simul 11
19. Di Persio L, Honchar O (2016) Artificial neural networks approach to the forecast of stock market price movements. Int J Econ Manage Syst 1
20. Eapen J, Bein D, Verma A (2019) Novel deep learning model with CNN and bi-directional LSTM for improved stock market index prediction. In: 2019 IEEE 9th annual computing and communication workshop and conference (CCWC), pp 0264–0270
21. Wen M, Li P, Zhang L, Chen Y (2019) Stock market trend prediction using high-order information of time series. IEEE Access 7:28299–28308
22. Yeoh W, Jhang YJ, Kuo SY, Chou YH (2018) Automatic stock trading system combined with short selling using moving average and GQTS algorithm. In: 2018 IEEE international conference on systems, man, and cybernetics (SMC), pp 1570–1575
23. Vargas MR, De Lima BS, Evsukoff AG (2017) Deep learning for stock market prediction from financial news articles. In: 2017 IEEE international conference on computational intelligence and virtual environments for measurement systems and applications (CIVEMSA), pp 60–65
24. Jozefowicz R, Zaremba W, Sutskever I (2015) An empirical exploration of recurrent network architectures. In: International conference on machine learning, pp 2342–2350
25. Chung J, Gulcehre C, Cho K, Bengio Y (2014) Empirical evaluation of gated recurrent neural networks on sequence modeling. 2014. arXiv preprint arXiv:1412.3555
26. Yahoo finance—stock market live, quotes, business & finance news (2020 (Accessed May 1, 2020)). https://in.finance.yahoo.com/
27. Bhanja S, Das A (2018) Impact of data normalization on deep neural network for time series forecasting. arXiv preprint arXiv:1812.05519

Chapter 17
Analysis of Structure and Plots of Characters from Plays and Novels to Create Novel and Plot Data Bank (NPDB)

Jyotsna Kumar Mandal, Sumit Kumar Halder, and Ajay Kumar

Abstract This paper consists of a technique that helps to build a well-structured character interaction network from plays and novels. In addition, the resulting network describes the leading characters along with their gender. Moreover, the network analyzes the gender of the central characters to the plots preferred by the author. We propose a databank called Novel and Plot Data Bank (NPDB) to store the relevant information by computing informative properties of the resulting network.

Keywords Character network · Graph analysis · Novel and Plot Data Bank (NPDB) · Information retrieval

1 Introduction

A character network is a graph extracted from a text, which represents the relationship among characters, where nodes represent the characters and edges represent the interaction among the characters. Our focus is on qualitative research on play and novels. Qualitative research allows the researchers to examine the plot, structure, and character interactions of plays and novels for a final decision. A character network is helpful to understand the summary of plays and novels. It is time saving as we do not need to read and analyze a large number of texts. On the other hand, we need a data bank to store the relevant information of the resulting character network. In our

J. K. Mandal (✉)
Department of Computer Science and Engineering, Faculty of Engineering, Technology and Management, University of Kalyani, Kalyani, West Bengal, India
e-mail: jkmandal@klyuniv.ac.in

S. K. Halder
Department of Computer Science, Faculty of Computer Science, Mahadevananda Mahavidyalaya, West Bengal State University, Barasat, West Bengal, India

A. Kumar
BSNL Ranchi, Ranchi, India

J. K. Mandal et al. (eds.), *Proceedings of International Conference on Innovations in Software Architecture and Computational Systems*, Studies in Autonomic, Data-driven and Industrial Computing, https://doi.org/10.1007/978-981-16-4301-9_18

work, we propose a well-defined data bank called Novel and Plot Data Bank (NPDB) to store the structural data of a character network.

We divided the entire work into three main modules: (1) Build a well-structured character interaction network from plays and novels, (2) compute the resulting network properties, and (3) build the Novel and Plot Data Bank. In this paper, we analyze ten of Rabindranath Tagore's plays and novels. We discuss all three modules, along with data sources, in more detail below. First, we briefly study some related work.

2 Related Work

The plot of a novel is defined by the group of its characters and the actions they carried out. Each novel is concerned about its impactful characters and their quantification in the plot. And the plot altered accordingly within the structure of the characters in a novel. In 2012, Graham Sack [1] proposed the mechanism, 'character networks for a narrative generation' by using social networks for producing plots artificially. One of the primary efforts was in 2002 by Alberich et al. [2] of linking social networks and literature. The resulting network was very comparable to a real social network that was based on Marvel Comics, where characters are the nodes associated with co-occurrence within the same book. To discover groups of characters, Newman and Girvan [3] in 2003 used a hand-build social network with a core character of Victor Hugo's Les Miśerables that noticeably reflected the subplot structure of the book. In 2010, Elson et al. [4] proposed a technique where characters are associated if they converse. The networks are constructed instinctively, and heuristics are used to group co-referents, which deflates extensive-standing literary premises. To evaluate semantic directions of social networks from the literature, Celikyilmaz et al. [5] offered extraction of dialogue interactions in 2010. Researchers like Rydberg-Cox [6] in 2011, Suen et al. [7] in 2013 and Giorgino [8] in 2004 extract networks from structured documents and corresponding text classification is done to implement large-scale studies. Genre-based clustering, openion mining, exploring of network structure and open source implementation of the analysis of plot structure are also done in recent years [9–13].

All the approaches mentioned above produced static networks. The validity of static network analysis is poor because of its flat illustration of the novel. In 2012, Agarwal et al. introduce the concept of dynamic network analysis for the literature which is a group of independent networks for each of the parts within which the novel is distributed. In the most recent time, Vincent and Xavier [14] in 2019 done a survey extraction and analysis of fictional character networks. Section 3 of the paper deals with proposed methodology with results of implementations. Conclusions are drawn in Sect. 4. References are given at end of the paper.

3 Proposed Methodology with Results

The proposed work can be divided into six modules. They are data collection, document preprocessing, character name recognition, gender assignation, network construction and analysis, and construction of the NPDB.

3.1 Data Collection

First of all, we need to collect all the necessary script files before preprocessing and analysis of the data. Scripts are freely available online. For our work, we downloaded the script files from tagoreweb.in and archive.org Web sites. We also required a text file that contains commonly used English words. The common file is also available online. We downloaded it from github.com Web site. The common file used the file to remove non-name words from the scripts while generating character names.

3.2 Document Preprocessing

After collecting all the relevant files, we need to perform preprocessing on the scripts. The purpose of the preprocessing is to improve the document efficiency for storing as well as retrieving the data. Some common document preprocessing steps are tokenization, case normalization, removal of punctuation, stop word elimination, stemming, lemmatization, etc.

3.3 Character Name Recognition

After the preprocessing of the text, we need to identify the characters of the script. To perform the task, tokenize the whole scripts into sentences. To classify the characters, we need to use Named Entity Recognition (NER). For our work, we used the pre-trained classifier called spacy NER. Then, we need to perform a character resolution process. This process improves the accuracy of the occurrence of a character in the script. For instance, 'Miss Sen,' 'Miss Charulata Sen,' 'Charu,' 'Charulata Sen,' and 'Charulata,' all of the names are corresponding to only one character of the 'Red Oleanders' play. After all the single name word created, we eliminate the names which are contained in the common word text file. To reduce the less important characters, we used a threshold value. In our case, we used ten as a threshold value. To identity important characters of a script, we counted the frequency of each character. This task can be done with the CountVectorizer () of the Scikit_Learn module.

Table 1 Gender assignation along with accuracy

S. No.	Play and novel	No. of leading character	No. of male character	No. of female character	Accuracy
1	Red Oleanders	10	9	1	1.00
2	Sacrifice	7	5	2	0.85
3	Chitra	4	2	2	1.00
4	Malini	7	4	3	0.85
5	The Post Office	8	7	1	1.00
6	The Trial	7	1	6	0.85
7	Debt Settlement	8	0	8	0.75
8	The King and the Queen	10	7	3	0.80
9	Autumn Festival	5	0	5	0.80
10	Sravanagatha	3	0	3	0.67

3.4 Gender Assignation

We took the gender name dataset from NLP's corpora and use it as training data for the Naive Bayes classifier that probabilistically classifies lists of names into a split of the genders by determining arrays within gendered names from tested data (Table 1).

3.5 Network Construction and Analysis

To visualize the graphical representation of the character network, we need to build a conversational network that is well suited for play and novels. In the graph, characters are represented by nodes and spoken interactions between characters are represented by edges. The graph is undirected due to the direction of the interactions ignored and also weighted as several interactions between the two linked nodes (Fig. 1).

3.6 Novel and Plat Data Bank (NPDB)

We need a database to store all the relevant information structurally. Mainly, NPDB database contains the characteristics of the resulting network. We store the information in a .csv file. The.csv file contains the following attributes:

- Node ID: Characters of a novel are represented by the Node ID.
- Chain ID: Connectivity or the pairs between two nodes are represented by the Chain ID.

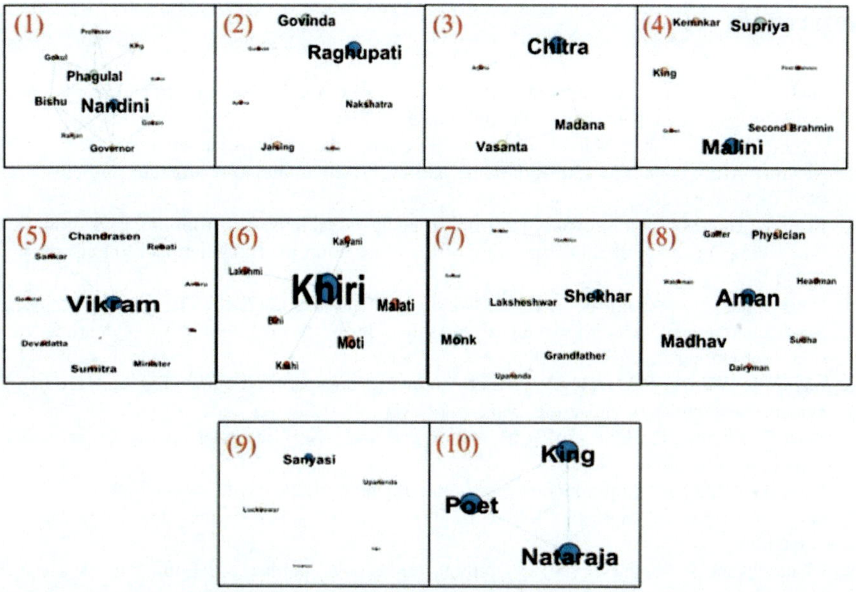

Fig. 1 Character networks of ten Tagore's plays and novels

- Sequence: Order of the appearance of characters of a novel is represented by the sequence.
- Label: Names of the character.
- Type: Directed or undirected graph.
- Weight: Edge values between two nodes.

4 Conclusions

In our work, we successfully extracted character networks from plays and novels. We compute informative properties from the resulting networks. We find out the leading characters along with their gender. The accuracy of the gender analysis in our study is 85.7% on average. We also proposed a databank called NPDB to store the relevant information about the resulting networks.

There are various operations that need to perform in the future. We have to focus more on NPDB architecture to build a compact data bank. We need to develop a Web server for analysis of the network and its component dynamically. The character network analysis can be extended for Web page script classification. In the future, the networks must be designed and implemented to classify a huge set of scripts to achieve improved performance to prove its effectiveness.

References

1. Sack G (2012) Character networks for narrative generation. In: Eighth artificial intelligence and interactive digital entertainment conference
2. Alberich R et al. (2002) Marvel universe looks almost like a real social network
3. Newman MEJ, Girvan M (2003) Finding and evaluating community structure in networks. Phys Rev E 69:1–16
4. Elson D, Dames N, McKeown K (2010) Extracting social networks from literary fiction. In: Proceedings of the 48th annual meeting of the association for computational linguistics, pp 138–147
5. Celikyilmaz A, Hakkani-tur D, He H, Kondrak G, Barbosa D (2010) The actortopic model for extracting social networks in literary narrative. In: NIPS workshop: machine learning for social computing
6. Layton R, Watters P, Dazeley R (2011) Automated unsupervised authorship analysis using evidence accumulation clustering. Nat Lang Eng 19:95–120
7. Suen C, Kuenzel L, Gil S (2013) Extraction and analysis of character interaction networks from plays and movies
8. Giorgino T (2004) An introduction to text classification. Retrieved on 13 Oct 2004
9. Gupta S, Becker H, Kaiser G, Stolfo S (2005) A genre-based clustering approach to content extraction
10. ChandraKala S, Sindhu C (2012) Opinion mining and sentiment classification: A survey. ICTACT J Soft Comput 3(1):420–425
11. Networkx.github.io (2020). Tutorial—NetworkX 2.4 documentation. [Online]. Available: https://networkx.github.io/documentation/stable/tutorial.html
12. Hagberg A, Swart P, Chult DS (2008) Exploring network structure, dynamics, and function using NetworkX. No. LA-UR-08-05495; LA-UR-08-5495. Los Alamos National Lab. (LANL), Los Alamos, NM, United States
13. Bastian M, Heymann S, Jacomy M (2009) Gephi: an open source software for exploring and manipulating networks. In: Third international AAAI conference on weblogs and social media
14. Vincent L, Xavier B (2019) Extraction and analysis of fictional character networks: a survey. ACM Comput Surv 52:89. https://doi.org/10.1145/3344548

Author Index

J. K. Mandal et al. (eds.), *Proceedings of International Conference on Innovations in Software Architecture and Computational Systems*, Studies in Autonomic, Data-driven and Industrial Computing, https://doi.org/10.1007/978-981-16-4301-9